T0133333

Imran Muhammad

Virtual Building Acoustics

Auralization with Contextual and Interactive Features

Logos Verlag Berlin GmbH

λογος

Aachener Beiträge zur Akustik

Editors:
Prof. Dr.-Ing. Janina Fels
Prof. Dr. rer. nat. Michael Vorländer
Institute for Hearing Technology and Acoustics
RWTH Aachen University
52056 Aachen
www.akustik.rwth-aachen.de

Bibliographic information published by the Deutsche Nationalbibliothek

The Deutsche Nationalbibliothek lists this publication in the Deutsche Nationalbibliografie;
detailed bibliographic data are available in the Internet at http://dnb.d-nb.de.

D 82 (Diss. RWTH Aachen University, 2022)

Logos Verlag Berlin GmbH 2022

ISBN 978-3-8325-5601-3

ISSN 2512-6008

Vol. 38

Logos Verlag Berlin GmbH
Georg-Knorr-Str. 4, Geb. 10,
D-12681 Berlin

Tel.:   +49 (0)30 / 42 85 10 90
Fax: +49 (0)30 / 42 85 10 92
http://www.logos-verlag.de

Contents

Chapter 4: Sound Insulation Filters: Auralization

Chapter 8: Summary

Chapter 9: Outlook

Annexes

Bibliography

Curriculum Vitae

Abstract

Building acoustic auralization is used to assess the perceptual aspects of sound transmission in built environments to provide the guidelines for architectural constructions and to evaluate the noise effects on humans. These noise effects have a negative influence on daily-life activities and create disturbances in physical and mental work. These disturbances are present within the dwellings and/or might be from outdoor moving transient sound sources. Extensive research is carried out to predict the sound propagation and transmission in the buildings. Methods are available for auralization of sound insulation between connected rooms in compliance with the standardized data formats of sound insulation metrics and building structural geometries. However, there still exist certain challenges to be addressed to construct the transfer functions between noise sources and receiver rooms for indoor situations as well as for the outdoor moving sound sources and to make these sources audible through audio-visual virtual reality systems in real time and interactively. These challenges are because of certain simplifications which are implicit in the formulations on which the sound insulation prediction models are based, such as, diffuse field assumptions, neglect of source characteristics, and source and receiver room acoustics.

This thesis focuses on addressing the present challenges in the traditional sound insulation rendering techniques and establishing an interface between psychoacoustic researches and building acoustics in dwellings (especially airborne sound insulation) integrated with audio-visual virtual reality environments. From technical perspective improvements are made in sound insulation prediction methods, and corresponding filter construction and rendering techniques for auralization. In the first place, the building elements are considered as multitude of secondary sources rather than taking them as point source radiators and the bending wave patterns are addressed in order to be able to properly construct the transfer functions from source to the receiver room. Secondly, the room acoustical simulations are carried out for both source and receiving rooms to generate transfer functions from source to the source room walls and from radiating receiving room walls to the listener, so that the geometries and absorptions might be fit to the properties desired by the user for the spatial impression of the listening rooms. In addition, the transfer functions from radiating walls of the receiving room to listener are designed in such a way that not only indoor sources are handled nevertheless the outdoor moving sources are also addressed.

On the other hand, some important conditions are associated with virtual building acoustics auralization research platform for advanced studies of noise effects in dwellings which are addressed in this thesis. The audio files, generally, used in listening tests are arbitrarily manipulated by audio samples without the background of a physical model of the built environments. They must comply with the standardized data formats of sound reduction indices (level differences) and/or sound transmission coefficients. Other data related to building structure such as geometry shall be strictly connected to the architectural design, the building materials and constructions. Otherwise the conclusions of the psychoacoustic experiments have no direct correlation with the architectural design, especially when presented through virtual environments. To achieve this, the auralization framework is extended toward real-time interactive audio-visual technology, i.e. VR technology, in order to be able to introduce more realism and, hence, contextual features into psychoacoustic experiments. The building acoustics framework is validated by taking indoor and outdoor example case studies. Listening experiments close to real-life situations are carried out by using this framework to show that this framework can be used as an alternate to design new test paradigms which help to better analyse and interpret the noise impact in building situations depending on the actual activity such as living, working, learning, and rest.

List of Symbols

A	Equivalent sound absorption area in the receiving room
A_o	Reference equivalent sound absorption area; for dwellings given as $10\ m^2$
B	Bending wave stiffness per unit length
c_o	Speed of sound in air $(= 340\ \frac{m}{s})$
c_B	Bending wave speed
c_L	Quasi-longitudinal wave speed
c_S	Shear wave speed
$\underline{D}$	Complex diffraction coefficient
D_{nT}	Standardized sound level difference
$D_{v,ij}$	Junction velocity level difference between excited element i and receiving element j
E	Young's modulus
E_{dir}	Direct energy in room
E_{rev}	Diffuse field energy in room
f	Frequency
f_c	Critical frequency of structural element
f_n	Modal frequency
f_{ref}	Reference frequency $(= 1000\ Hz)$
F	Transition function
G	Shear modulus

h	Thickness of a structural element
I	Sound intensity
I_i	Incident sound intensity
I_r	Radiated sound intensity
i, j	Indices for an element; for a transmission path ij, i indicates an element in the source room ($= F, D$) and j an element in the receiving room ($= f, d$)
k_o	Wave number in air
k	Angular wave number
k_B	Bending wave number in plate
k_p	Wave number in plate
K_{ij}	Vibration reduction index for each transmission path ij over a junction
L_i	Sound intensity level
L_s	Sound pressure level in the source room
L_R	Sound pressure level in the receiver room
L_w	Sound power level
l_o	Reference length ($= 1m$)
l_{ij}	Common coupling length between element i and element j
m	Mass per unit area of an element
m_o	Reference mass per unit area $\left(= 1\frac{kg}{m^2}\right)$
P_a	Sound power source
p	Sound pressure
p_s	Sound pressure in source room
p_R	Sound pressure in receiving room
Q_s	Source directivity

Q_j	Secondary source directivity
r_{rev}	Direct to reverb distance in rooms
R	Sound reduction index of an element
R_o	Sound reduction index with mass law
R'	Apparent sound reduction index
R_i	Sound reduction index for element i in source room
R_j	Sound reduction index for element j in receiving room
R_{ij}	Flanking sound reduction index
R_f	Sound reduction index of forced Transmission
R_{Dd}	Sound reduction index for airborne direct transmission
R_{Ff}	Sound reduction index for the transmission path Ff
R_{Df}	Sound reduction index for the transmission path Df
R_{Fd}	Sound reduction index for the transmission path Fd
S_D	Area of separating element (partition) between adjacent room
S_i	Area of an element i in source room
S_j	Area of an element j in receiver room
t	Time
T	Reverberation time in the receiving room
T_o	Reference reverberation time; for dwellings given as ($= 0.5s$)
T_s	Structural reverberation time of an element
$\langle \hat{u}^2 \rangle$	Mean square velocity amplitude taken over the surface
V	Volume of the receiving room
v	Sound particle velocity
ν	Poisson's ratio

W_{ij}	Radiated sound power by element j due to incident sound on element i
W_s	Sound power incident on a test specimen in the source room
Z_w	Wall impedance
Z_0	Impedance of air
α_i	Equivalent absorption length of a structural element i
α	Absorption coefficient
α_m	Equivalent absorption area of a structural element
η_{ij}	Coupling loss factor between element i and j
η_{tot}	Total loss factor
η_{int}	Internal loss factor
θ	Angle of incidence
θ_c	Coincidence angle
ξ	Particle displacement
ρ	Density of structural element
ρ_0	Density of air
$\sigma(\theta)$	Angle dependent radiation factor
σ_r	Radiation factor for resonant transmission
σ_f	Radiation factor for forced waves
τ'	Apparent transmission factor
$\tau(\theta)$	Angle dependent transmission factor
τ	Transmission factor
τ_{ij}	Flanking transmission factor
τ_d	Flanking transmission factor to direct element $(Dd + Fd)$
τ_f	Flanking transmission factor to direct element $(Df + Ff)$

τ_{Dd} Flanking transmission factor from direct to direct element (separating element)

τ_{Ff} Flanking transmission factor from flanking to flanking element

τ_{Df} Flanking transmission factor from direct to flanking element

τ_{Fd} Flanking transmission factor from flanking to direct element

τ_e Transmission factor for portal energy radiations (doors, windows etc.)

τ_s Transmission factor of indirect airborne energy radiation

ψ Azimuth angle for k_p

ω Angular frequency

List of Figures

List of Tables

Annexes

1

Introduction

The ability to predict sound and vibration transmissions in structures such as buildings, trains and automobiles is important for assessment of human comfort, health and safety in planning processes. In modern societies there is a concern about steadily growing annoyance due to the indoor and outdoor noise in private dwellings as well as in commercial worksites. Background speech in commercial worksites leads towards reduced power of concentration during physical or mental work [1]. This is, of course, due to increase in traffic noise and excessive usage of the electrical and mechanical utilities. Different surveys reveal that in multifamily apartments people are annoyed by noise mainly caused by indoor activities [2,4]. The studies also show that people are exposed to the noise from neighbours, which causes consequences of disturbance in sleep, physical or mental work impairment, and the disturbance in conversation or listening to the TV or radio in private dwellings as well as working performance in office premises. On the other hand, intelligible background speech such as conversation or phone call, is considered as a negative feature of office environment [5]. While the noise is steadily increasing in densely populated urban areas, building structures, and corresponding guidelines and standards of sound insulation requirements are still very similar to those decades ago.

In this regard in residential and worksite premises, especially in urban areas, the international standards provided by ISO (International Standards Organization) have reflected the trends mentioned above increasing both in number and covering broader aspects. On the European stage, sound insulation guidelines, such as ISO [6,17,69] and DIN-4109 [28], are provided by the government to protect citizens from the noise exposures. However, these guidelines do not provide an optimal acoustic satisfaction especially when specific sounds, for example a conversation varying in intelligibility or an intermittent noise, originate either from the adjacent office or from facades causing the disturbances in daily life work performances.

Therefore, measurement procedures applied in laboratory or in the field are just one part of the story. There are numerous guidelines and standards that describe the performance of building elements in terms of general sound level reduction indices in the form of a single number value and/or frequency dependent curves. However, it can be assumed that these quantities are insufficient to describe the specific situation for the perception of noise [7]. The performance of the buildings concerning protection against noise can be evaluated from a technical perspective as well as in a human-centred approach by considering subjective ratings, cognitive performances, or other human activities [8]. The technical-oriented evaluation is based on the standard measurements and prediction techniques. Laboratory and field measurements as well as prediction models for planning purposes are quite advanced. The evaluation and single-number rating, i.e., the so-called "sound insulation metrics" are used in practice to ensure a proper noise protection, for which limits (requirements) are set by national authorities (e.g. ISO-12354; ISO-717; ISO-140). The measurements, predictions and decisions about noise control in building acoustics are based on the correlation of the sound insulation metrics with subjective ratings and with field surveys at the national level [4] with different metrics and different noise limits. Therefore, it is an appropriate to use simulation tools which estimate the sound field at the ears of the listener from the predicted or measured data using auralization of these sounds which better consider the subjective impression of the subjects, and psychoacoustic or psychological factors.

1.1. Background and Related Work

The basic principle of building acoustic auralization is to simulate the alteration of a sound signal from its source to the receiver via transmission through the building structures [8]. The auralization of an office-to-office situation requires modelling of sound propagation in both office rooms, where the sound from one office is transmitted through building structures to another neighbouring office. On the other hand, for an outdoor moving sound source, the modelling of outdoor sound field and its transmission through façade elements of buildings is involved. Sound propagation modelling involves its generation, transmission form building elements and the insulation characteristics of the direct and flanking elements between source and receiver room [9]. Both level and spectral characteristics of the transmitted noise highly depend on the insulation curves of the building constructions separating source and receiver [3,10,14]. Several methods are available for auralization of sound insulation in compliance with the standardized data formats of sound insulation which calculate transfer functions from noise source to the listener end. However, there are certain challenges to be addressed in order to design the sound insulation transfer functions between source and receiving rooms to achieve plausible auralization. These challenges are because of the assumptions on which the formulation and derivation of these existing insulation prediction models and sound transmission techniques are based. Two approaches are common in predicting sound and vibration transmission through building structures. At low frequencies the numerical methods, such as Finite Element Method (FEM) [11] or semi-analytic methods [12] may provide a quick and efficient calculations of the structural response.

These models, however, require computation times which exceed the limits of real-time processing (~50ms) by orders of magnitude [9]. For this reason, statistical approaches are used, such as Statistical Energy Analysis (SEA) [26] and Advanced Statistical Energy Analysis (ASEA). These methods are used for calculating the energy exchange between adjacent building elements and the respective energy losses, under steady-state conditions [13]. SEA models predict the average response of ensemble elements of the system, therefore, the coupling loss factors and modal densities represent ensemble average. On the other hand, the international standards [6,16,17], for example, are commonly used as guidelines for building constructions for prediction of sound insulation as first order SEA models.

Based on existing sound insulation prediction methods (e.g. ISO), the first application of auralization of airborne sound insulation was introduced by Vorländer and Thaden [3,9,10,14] who transferred Gerretsen's prediction method [18] into the signal and filter domain. They presented an auralization framework for sound insulation through binaural reproduction technology. Subsequently, these filters were used to calculate airborne sound transmission paths from source to receiver placed in simple adjacent rooms of the building. However, several simplifications were made in their approach. At first, it was assumed that the incident sound intensity on the source room elements (i.e. source room walls) is equal for all transmission paths. In other words, the same amount of incident sound intensity impacts on all source room elements independent of the source position, its directivity and complex room geometries. Secondly, the calculated transfer functions between source and receiver rooms were only valid for arbitrary point to point transmission [3]. Extended walls, however, are always present in real situations. To include the room reverberation, they used measured room impulse responses (RIRs). In the receiving room, the simplification was made that the sound is apparently radiating from one point located at the centre of the room element representing the whole bending wave pattern on the walls. The output of their work was also implemented in a commercial version called "BASTIAN" [**72**] that calculates airborne and impact sound transmission between adjacent rooms, and airborne sound transmission from the exterior sound sources as well. This software is based on measurement databases and ISO [**6**] for prediction of airborne sound transmission with offline auralization features. On other hand, the real-time room acoustic simulation software "RAVEN" developed by Institute of Technical Acoustics (ITA), RWTH Aachen University, relies on the knowledge of room acoustical simulation techniques and enables a physically accurate auralization of sound propagation in complex rooms, including important wave effects such as sound scattering, and sound insulation between rooms as well. In [**54**], the author introduced the concept of portals (i.e. doors, windows etc. between the connected rooms) and the receiving room walls as secondary sources (SS). However, sound insulation filters part in RAVEN is based on the work by [**3**] with the same building acoustic limitations, though the sophisticated room-acoustical simulation can be carried out with more accuracy and precision.

1.2. Research Objectives

As discussed, the recent up-to-date sound insulation prediction and auralization methods include several simplifications that are implicit in the formulation on which they are based. Therefore, there is an opportunity to develop a novel building acoustics auralization framework based on detailed models of ISO standards and available measurements, integrated with virtual reality (VR) systems, to accurately realize the perception and evaluation of noise and its influences on the humans. It may help to further develop guidelines for building constructions. One advanced goal of insulation auralization in real time ("virtual building acoustics") is to appropriately reproduce the condition of noise effects on the human perception and cognitive performance. These effects depend on the kind of noise signal (speech, music, traffic noise etc.) and on the context [**19**]. This way, studies on sound perception can be performed in a more ecologically valid approach since in real-time VR the user can freely move, turn the head, or change scene parameters during runtime. This is a significant difference to the state-of-the-art audio demonstrations where the sound is played to the user without any of these adaptation of interactive components.

From the technical point of view, the aim is to develop real-time airborne sound insulation rendering techniques for auralization of virtual built environments (e.g. private dwellings, and commercial office sites). The objective of this work is to design airborne sound insulation filters for sound transmission between adjacent rooms separated by building elements and for the outdoor sound sources passing by the buildings. The methods and the approaches are designed in such a way that they use ISO standard [**6,17**] as building blocks for airborne sound prediction and more focus is on addressing the simplifications that exist in the previous techniques [**3,7,10,14**]. On the other hand, the human-centred approach is adopted for dynamic building acoustics auralization and perception to realize a real-time audio-visual virtual reality framework for psychoacoustic and psychological experiments, and as a demonstrator. Therefore, an accurate airborne sound transmission filter design strategies and auditory visual virtual reality framework is targeted in this work to represent a perceptually plausible building acoustics auralization in virtual reality.

In the first place, the reverberation of both source and receiving rooms is an important acoustical parameter which is taken into account depending on the room characteristics (e.g. room geometries, wall absorptions etc.), source directivities and the spatial variation of sound field inside rooms. Secondly, the sound insulation transfer functions from source to receiving rooms are calculated for extended walls by using concept of segmenting individual building element into a multitude of secondary sound sources with non-uniform energy distribution by considering running bending waves patterns on the flanking elements according to the energy decay in structures. Furthermore, the directivity patterns of the radiating elements were so far not addressed, as the radiation efficiencies of the vibrating plates play an important role in sound field propagations inside the rooms. The flanking elements may be homogeneous (e.g. a single homogeneous wall element) or consisting of an assembly of two or more parts or surfaces (e.g. doors, windows). Thirdly, the receiver room acoustics is implemented in a way that it includes the receiving room reverberation based on room geometry, absorption and binaural transfer functions between radiating walls and the receiver. In this way, it is possible to experience binaural sound from arbitrarily positioned sound sources inside the source room to a dynamically moving listener in the receiving room. Moreover, the room impulse responses (RIRs) are synthesized from one-third octave band values of the reverberation times of source and the receiving rooms based on the proposed method in [**64**]. Finally, the algorithms are implemented and tested in virtual environment as a virtual reality application for subjective evaluation of sound insulation in building environments. This framework is a real-time auditory-visual virtual reality framework and verified and validated while conducted listening experiments regarding the evaluation of noise effect patterns and background speech effects on human cognitive performance under various building acoustical conditions for adjacent rooms as well as for façade sound insulation against outdoor moving sound sources. These experiments, on the one hand, explore the impact of background speech differing in intelligibility and level on verbal serial recall and thus verbal short-term memory capacity and on the other hand explore the localization capabilities of the outdoor moving sound source under façade sound insulation conditions. The first experiment was simple; e.g. static source and receiver in both source and receiver rooms, however, in everyday life the dynamic scenes are present where the people perform tasks of daily life of work or learning under conditions of usual behaviour and movement. In the second experiment, i.e. the outdoor moving sources, the impact of intermittent noise effects of passing-by car, ambulance and police sirens are investigated, which is of more interest for defining sound insulation guidelines and the regulation for acoustic comfort regarding urban environments. Hence, the outcomes and the results of the presented work would increase interaction with virtual environments, making a more realistic and immersive scene and leads toward an accurate subjective evaluation of building performance. This framework of building acoustic auralization can be used in listening experiments and allows the test subjects to perform any task of daily life of work or learning under conditions of usual behaviour and movement. Therefore, it can create more realistic noise perception tests in interactive real-time virtual reality environments.

1.3. Content Outline

The thesis is organized in nine chapters starting from introduction, background and research outlines. Chapter 2 introduces the fundamental concepts and theory behind the sound propagation and building acoustics concepts. Specific background knowledge of room acoustics is provided in this chapter to facilitate the reader in understanding the basic concepts of room acoustics quantities which are used for the computations of the building acoustic parameters. Furthermore, this chapter discusses the fundamentals of sound transmission and sound radiations through building structures and theories for the calculation of sound insulation for building elements. Chapter 3 starts with introducing airborne sound transmission through a variety of the building elements, such as finite plate and infinite plate structures. Different sound transmissions, such as direct sound transmission and flanking transmission, are elaborated with derivations of the transfer functions for sound insulation. Classical methods for sound insulation predictions are discussed with their limitations as well as extended approaches are introduced for advanced airborne sound insulation prediction. Chapter 4 reviews the building-acoustic fundamentals and the concepts of indoor and outdoor sound fields. Sound transmissions through the building structures, radiations form the wall elements and their types are discussed to use in filter design process. Chapter 4 further extends the knowledge of airborne sound insulation filter and transmission by dwelling by direct as well as by flanking paths and describes the fundamental auralization techniques based on developed sound insulation models and available standards that are used for the final binaural reproduction at the listener end in the receiver room. The filter development process is explained for adjacent rooms as well as for the outdoor sources. The corresponding real-time algorithm techniques are presented and their properties are examined in detail.

In Chapter 5, the developed sound insulation models are implemented and verified with standards and available measurement data. Different case studies are taken into account, such as adjacent office-room case (for indoor) and urban cases (for outdoor), for the verification of final outcomes. These outcome are the level differences D_{nT} between the source and the receiving rooms. Chapters 6 introduces the applications of developed framework in audio-visual virtual reality environments. Applications of the framework are discussed focusing on the evaluation of performance of buildings and on the design of particular listening experiments; such as, evaluation of background noise impacts on the cognitive performance of humans under different building acoustical conditions and on effects of intermittent outdoor moving sound sources on perceptual localization capabilities of the human. Furthermore, example studies for listening experiments are described which allow the test subjects to perform any task of daily life of work or learning under conditions of usual behaviour and movement. Last but not least, Chapter 7 begins with the aim that building acoustic auralization framework in virtual reality can better serve as a valuable tool to assess the perceptual aspects of sound transmission in built-up environments. In this chapter it is shown that different psychoacoustic experiments are possible in virtual reality for evaluation of noise in virtual building environments and psychological research can be conducted in an ecologically more correct way. Finally, towards the end of this thesis, the findings are summarized, and open scientific questions are referred in the outlook with the future work and challenges in the presented research.

2

Fundamentals of Building Acoustics

Building acoustics is a research area that covers all aspects of sound and vibration in the built environments. This chapter describes the basic concepts behind the acoustical performance of the buildings as derived from the performance of the elements, comprising various structures. The fundamentals of sound propagation in the closed spaces (i.e. room acoustics) are discussed and explored, as most of the building acoustic theory for indoor environments is based on spatial and temporal aspects of sound propagation inside rooms (e.g. the modal sound field and the diffuse sound field). Furthermore, a brief introduction of outdoor sound propagation in the urban areas is discussed to understand the airborne sound insulation against outdoor source (e.g. façade sound insulation). Nevertheless, the main emphasis is on real-time simulations of sound insulation, sound transmission, and the basic aspects of noise problems connected with internal and external noise sources. External noise sources are major characteristics of urban environments which are important regarding the human comfort either caused by transportation, industrial facilities, civil constructions or recreational and residential activities. These are integral part of the daily life of urban inhabitants even for insulated indoor spaces, such as, dwellings, workspaces and classrooms. Thus, the physical aspects of sound transmission need to be understood and the focus in this chapter is to discuss these physical aspects during design of sound insulation filters in order to provide the basis for an interactive auralization framework.

This chapter starts with the room acoustic concepts that include all aspects of the behaviours of the sound field in rooms, covering both the physical aspects as well as the subjective effects. In other words, room acoustics deal with measurement and prediction of the sound field resulting from a given distribution of sources as well as how a listener experiences this sound field [9,21]. This implies that having a knowledge on how the shape of the room, the dimensions and the material properties of the construction influences the sound field. The reverberation in a room is an important characteristic in any judgement of properties of the sound field on which the building acoustic prediction models are based. There are other parameters that are based on the impulse response of the room but here the emphasis is on the relative energy contents in the given time intervals. From this knowledge of room acoustics, we may derive the information about sound field hitting the surfaces of the building elements (i.e. from the adjacent room) and start our expedition toward sound transmission of this sound field through building structures and prediction of sound insulation metrics. Moreover, the basics of outdoor sound propagation in urban areas are discussed considering the direct sound field and early reflections from the surrounding building facades. An important aspect of building acoustics is frequency range which generally is considered in the range defined by one-third octave bands from 50 Hz to 5000 Hz. Airborne sound insulation tends to be weakest in the low frequency range and highest in the high-frequency range which means that it is acting as low-pass filter. Hence, significant transmission of airborne sound above 5000 Hz is not usually an issue. However, when auralization of sound insulation comes into play, it very important to consider sound insulation filter design at frequencies outside the building acoustics frequency range in signal and system domain. Therefore, interpolation and extrapolation techniques are applied very carefully to cover a full audible frequency range which is typically from 20 Hz to 20,000 Hz.

2.1. Sound Field in Rooms

The sound fields in rooms are of the primary importance in studying sound insulation and prediction. This section describes the basic principles of sound fields in different aspects in more detail that is relevant to the prediction of sound insulation metrics. For the building acoustics frequency range, the sound field is often considered as diffuse field which is a very useful and idealized concept for prediction of sound insulation and sound reduction index. A diffuse field composed of a large number of statistically independent plane waves, the spatial phase of which is uniformly distributed and independent from the amplitude. As described in [78], the diffuse field may represent the sound field of a conceptual ensemble of rooms with the same modal density and total absorption, however, any possible arrangement of boundaries and small objects that scatter incoming sound waves. For example, if we choose any point in the closed space, sound waves arriving at this point have random phases and there is an equal probability of sound waves arriving from any direction. In real situations there is a wide variety of rooms with different sound fields prevailing in such spaces. These fields, generally, are interpreted with reference to two idealized models, which are the modal sound fields and the diffuse sound fields, however, the diffuse field is just a concept. Adopting a diffuse field model therefore inherently implies that uncertainty due to random wave scattering is present in the computed results. This uncertainty can be large, especially at low frequencies [78]. In practice there is always a dissipation of energy, therefore, there may not be an equal energy flow in all directions. There must be a net energy flow from sound source towards the part(s) of a closed space where sound is absorbed. In the diffuse field it is common to refer to diffuse reflections which means that the relationship between the angle of incidence and the angle of reflection is random [21]. The diffuse sound field is a useful concept that allows many simplifications which might be assumed, during the measurements, for the prediction of sound insulation and other room acoustic calculations. However, the sound field does not always bare a close resemblance to the diffuse sound field over the entire building acoustics frequency range [22]. In the low-frequency range this is primarily because sound waves can be decomposed into a relatively low number of modes arriving at any point and corresponding to a limited number of directions, whereas, in the mid and high frequency ranges the sound waves arriving at any point tend to come from many different directions. It means that the assumption of diffuse sound field at low-frequencies may not be a suitable choice to produce realistic prediction of sound insulation.

2.1.1. Direct and Diffuse Fields, Reverberation Distance

The results of the stationary conditions and of sound decay in a room can be applied to measure the sound power, W_a, of a sound source as given in Equation **2.1** with p_{diff}^2 representing the diffuse sound pressure, V denoting the volume and T is the reverberation time of the room [21]. Here, c_o and ρ_o are the speed of sound in the air and air density respectively.

$$W_a = \frac{p_{diff}^2}{4\rho_0 c_0} \cdot \frac{55.3 \cdot V}{c_0 T} \tag{2.1}$$

In ordinary rooms, the diffuse field is generally a rather simple approximation to the stationary sound fields. It is used to separate direct sound from the reverberant part of the sound field [21,22]. The sound power radiated by an omnidirectional source is the sound intensity at a distance r in a spherical field multiplied by the surface area of a sphere and is the direct sound which is given in Equation **2.2** as,

$$p_{dir}^2 = \frac{\rho_0 c_0 W_a}{4\pi r^2} \tag{2.2}$$

and the stationary sound (i.e. diffuse filed) in Equation **2.3** as,

$$p_{diff}^2 = \frac{4\rho_0 c_0 W_a}{A} \tag{2.3}$$

The reverberation distance r_{rev} is defined as the distance where $p_{dir}^2 = p_{diff}^2$ when an omnidirectional point source is placed in a room. It is a descriptor of the amount of absorption in a room since the reverberation distance depends only on the equivalent absorption area A, therefore, is given in the form of Equation **2.4**.

$$r_{rev} = \sqrt{\frac{A}{16\pi}} = 0.14\sqrt{A} \tag{2.4}$$

The physical interpretation of reverberation distance implies that at a distance closer to sound source than the reverberation distance direct sound field dominates which is named as direct sound field, whereas, at larger distances to sound source the reverberant sound field dominates. In this so-called "far field", the diffuse field sound pressure may be a useful approximation. An expression for the combined direct and diffuse sound field is derived by simple addition of the squared sound pressures of two sound fields. For this, the sound power of the source should be reduced by a factor of $(1 - \alpha_m)$, which is the fraction of the sound power emitted to the room after the first reflection [**22**]. The total squared sound pressure in then,

$$p_{total}^2 = p_{dir}^2 + p_{diff}^2(1 - \alpha_m) = p_{diff}^2 \left(\frac{r_{rev}^2}{r^2} + 1 - \alpha_m \right) \tag{2.5}$$

$$p_{total}^2 = W_a \rho_0 c_0 \left(\frac{1}{4\pi r^2} + \frac{4}{A}(1 - \alpha_m) \right) \tag{2.6}$$

The absorption area A divided by $(1 - \alpha_m)$ is called the "*room constant*". Typical sound sources such as speaking persons, loudspeakers or musical instruments radiate sound with different intensities in different directions. Their directivity factor

Q, which is the ratio of the intensity in a certain direction to the average intensity, is defined in Equation **2.7**, with I as the sound intensity.

$$Q = \frac{4\pi r^2 I}{W_a} \tag{2.7}$$

Which makes the squared sound pressure of the direct sound equals to $p_{dir}^2 = \frac{\rho_0 c_0 W_a Q}{4\pi r^2}$, and leads to a general formula for the sound pressure level as a function of directivity and distance from the sound source in the room and is given in Equation **2.8**. (Here $A_0 = 1m^2$ is taken as per ISO Standards).

$$L_p \cong L_w + 10\lg\frac{4A_0}{A} + 10\lg\left(Q\frac{r_{rev}^2}{r^2} + 1 - \alpha_m\right) \tag{2.8}$$

As described in [22], in case of omnidirectional sound sources and in a reverberant room with small sound absorption ($\alpha_m < 0.1$), the sound level in the far field is approximately predicted by diffuse filed theory. It means that the last term in Equation **2.8** will be close to zero. Whereas, in case of highly directional sound sources (i.e. trumpet), the direct field is extended to distances much longer than the reverberation distance. Which means that the last term in Equation **2.8** now raises the sound pressure level above the diffuse field value. In relatively large rooms with medium or high sound absorptions, the sound pressure level continues to decrease as a function of the distance because the diffuse field theory is not valid in such rooms [22]. Instead, the slope of the spatial decay curve may be taken as a measure of the degree of acoustic attenuation in these rooms. A reverberant room can be used to determine the sound power of a source by measuring the average sound pressure level in the room. If measurements are made in positions that avoids the direct sound, the last term in Equation **2.8** becomes more correct [44], by an approximation: $10\lg(1 - \alpha_m) \rightarrow 10\lg\left(e^{\frac{-A}{S}}\right) \cong -4.34\frac{A}{S}$, with A as equivalent absorption area and S is the total surface area. However, at low frequencies to produce correct results the expression for the sound pressure level is determined by Equation **2.9**, which is taking into account the "Waterhouse correction" [83] and gives a standard measurement procedure for sound power level in a room [adopted by **ISO 3741**].

$$L_w = L_p + 10\lg\frac{A}{4A_0} + 4.34\frac{A}{S} + 10\lg(1 + \frac{c_0 S}{8Vf}) \tag{2.9}$$

2.1.2.   Incident Sound Power on a Surface

The relationship between the root-mean-squared (RMS) sound pressure, and sound intensity, I, for a plane propagating sound wave is given by

$$p^2 = I \rho_0 c_0 \tag{2.10}$$

In a diffuse sound field, the sound waves are propagating in all directions and hence the RMS sound pressure p_{diff} is the result of sound waves propagating in a diffuse sound field. By integration over a sphere with the solid angle $\psi = 4\pi$, diffuse sound pressure is given as [22].

$$p_{diff}^2 = \int_{\psi = 4\pi} I \rho_0 c_0 \, d\psi = 4\pi I \rho_0 c_0 \tag{2.11}$$

In the case of a plane wave with the angle of incidence θ relative to the normal of the surface, the incident sound power per unit area on the surface is

$$I_\theta = I \cos \theta = \frac{p_{diff}^2}{4\pi \rho_0 c_0} \cos \theta \tag{2.12}$$

This is just the sound intensity in the plane propagating wave multiplied by the cosine of the incident angle, which is the projection of a unit area as seen from the angle of incidence. The total incident sound power per unit area is found by integration over all angles of incidence covering a half sphere in front of the surface. Which is four times less than in the case of a plane wave of normal incidence.

$$I_{inc} = \int_{\Psi = 4\pi} I_\theta d\Psi = \frac{p_{diff}^2}{4 \rho_0 c_0} \tag{2.13}$$

2.2. Outdoor Sound Fields

In urban environments, the sound sources and especially the intermittent sources are audible even though there is no direct sound path, i.e. the direct line of sight between source and receiver. Outdoor sound fields propagate through different wave effects such as atmospheric absorption, scattering, reflections from the surrounding buildings and from the ground, and diffractions from the edges and corners of the buildings facades. Hence, the received sound field at receiver end is a combination of the direct sound field (if the source and receiver have direct line of sight), multiple reflection paths from the surrounding building geometries and the diffraction paths from the edges and the corners of building facades. Thus, calculation of such paths requires specific methods. The reflections, which are treated according to the Snell's law and are of great importance, are most likely causing changes in the sound field properties (i.e. amplitude and phases etc.) depending on the properties of the obstacle it collides with and reflects back. The nature of reflections, such as specular reflections and/or diffuse reflections is very important in the formation of sound field which in turn decides which assumptions should be taken into account while predicting the façade sound insulation. For example, ISO 12354 Part-3 [17] assumes diffuse sound field while predicting façade sound insulation metrics, however, this is not exactly the case in real urban environments.

Nevertheless, the incident and reflected sound pressures can be represented by $p_i = \hat{p}e^{j(\omega t - kx\cos\theta - ky\sin\theta)}$ and $p_r = R\hat{p}e^{j(\omega t - kx\cos\theta - ky\sin\theta)}$, where $R = |R|e^{-j\gamma}$ is the reflection coefficient (with k as wave number). The most commonly used way to represent R is given in Equation **2.14**, where Z is surface impedance on which the sound field strikes and Z_0 is the impedance of the atmospheric medium (i.e. air). The surface impedance can be taken as angle independent and in this case it is known as locally reacting surface. The reflection coefficients can be represented by sound intensity as $|R|^2 = \frac{I_r}{I_i}$, with I_r as incident and I_i as reflected intensities of the sound field, provided that the intensities are considered as transported energies.

$$R = \frac{Z\cos\theta - Z_0}{Z\cos\theta + Z_0} \tag{2.14}$$

Diffraction occurs at the corners or edges of surfaces and is a wave effect which bends the sound wave into the geometrical shadow region allowing us to hear sounds around corners. This is the reason why the sound does not change abruptly after the direct line of sight is blocked. Diffraction is negligible if the dimensions of surface is small compared to the wavelength of the incident sound field, however, since the audible wavelength range is approximately between **17mm** and **17m**, it is of great importance for an urban environment to take diffraction phenomenon into account. The Huygens-Fresnel principle can satisfactorily describe a larger number of diffraction configurations where every point on a primary wave front can be thought of as a continuous emitter of secondary sources, producing a new wave front in the direction of propagation. The Doppler effect is another well-known acoustic effect. It occurs in case of a moving source or a moving receiver causing a change of frequency at the receiver's end. During auralization, this effect should be considered to perceive plausible sound effects of moving vehicles.

2.3. Propagation of Sound in Plates

To understand the physics behind sound insulation, knowledge about airborne and structure-borne sounds and especially about wave phenomena in plates are essential. Two basic wave types are generally present in extended materials and therefore independent of each other. At a boundary, however, there is coupling and energy transfer from one wave type to the other. This is the reason for the consideration of combined waves in plates, rods and beams. In building acoustics, the following wave type are very important and play a key role in sound transmission through monolithic homogeneous or heterogeneous materials. These are longitudinal waves, shear waves and bending waves (also known as flexural waves). It is not intended to go into details of the types of the waves. In this section, however, it is briefly introduced into bending waves and their important characteristics. Figure **2.1**, shows different types of waves.

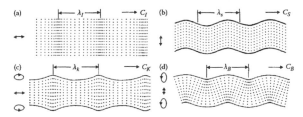

Figure 2.1: *Important types of waves. (a) Longitudinal; (b) Quasi-longitudinal; (c) Shear; (d) Bending [22]*

2.3.1. Longitudinal Waves

The ideal or pure longitudinal waves may only exist in a medium of infinite extent. Practically, this implies that the solid structure must be very large compared with the wavelength. When taking into account the actual dimensions of building elements and the relevant frequency range, displacements normal to the direction of wave propagation occur (i.e. longitudinal stresses will produce lateral strains on the outer free surfaces). This is called the Poisson contraction phenomenon. The associated wave type is therefore called quasi-longitudinal. Equation **2.15** shows that differential equation governing the free waves (one-dimensional case, i.e. the direction of propagation) in terms of $E' = \frac{E}{1-v^2}$ that is a property of the material which depends on the actual lateral displacement. A corresponding equation for the displacement could be used as well.

$$E' \frac{\partial^2 v_x}{\partial x^2} = \rho \frac{\partial^2 v_x}{\partial t^2} \tag{2.15}$$

The phase speed of the longitudinal wave, according to the Equations **2.15**, is given in Equation **2.16**. Here, E is Young's Modulus, v is Passion ratio.

$$c_L = \sqrt{\frac{E}{\rho(1-v^2)}} \tag{2.16}$$

2.3.2. Shear Waves

In a pure shear wave, also referred to as a transverse wave, we only get shear deformations and no change of volume. The particle movements are normal to the direction of wave propagation. The governing equation for shear waves is given as,

$$G \frac{\partial^2 v_y}{\partial x^2} = \rho \frac{\partial^2 v_y}{\partial t^2} \tag{2.17}$$

Where v_y represents the particle velocity normal to the direction of propagation. The shear modulus is given by

$$G = \frac{E}{2(1+v)} \tag{2.18}$$

And for the shear phase speed c_S, we get

$$c_S = \sqrt{\frac{G}{\rho}} \tag{2.19}$$

2.3.3.   Bending Waves (Flexural Waves)

Bending waves are likely to be excited in structures where one or two dimensions are small compared to the wavelength at an actual frequency. This implies that this wave type will be dominant in common construction elements (i.e. beams and plates). In other words, it takes a central position in building acoustics and also due to these waves it is easy to excite the structure [22]. Furthermore, the particle velocity will be normal to the direction of propagation, which also means that it is normal to the surface of a plate. It shows an efficient coupling to the surrounding medium (i.e. air), which means that the plate or beam potentially could be an efficient sound source. The general approach in building acoustic is that the treatment of bending waves is mainly restricted to the simple thin plate models, also called Bernoulli-Euler models. In these models, we can presuppose that the deformation of an element due to bending is much larger than the one caused by shear and, furthermore, the rotation of the element is neglected. A limit for using thin plate models, often referred in the literature, is that the wavelength of the bending wave must be larger than six times the thickness of the beam or plate. There is also another limitation that is implied on the treatment of the bending wave that is the plates are made of isotropic materials, (i.e. the material properties are independent of direction). In this case (i.e. isotropic materials), we need only two quantities, the modulus of elasticity and Poisson's ratio, to describe the linear relationship between forces and displacements. Unfortunately, a large group of building materials exhibit anisotropy such as wooden materials are typical examples. However, these materials are not the scope of this work. Therefore, we will restrict to treat the bending waves for homogeneous isotropic plates in the following section.

2.3.4.   Free Vibration of Plates

The differential equation for solutions of the bending wave are quite complex, therefore, we here just state that the equation for the particle velocity normal to the plate surface may be written in the form of Equation **2.20** with B as plate bending stiffness per unit length and m as mass per unit area of the plate.

$$-B\frac{\partial^4 v_y}{\partial x^4} = m\frac{\partial^2 v_y}{\partial t^2} \tag{2.20}$$

The same differential equation applies to other quantities such as displacement, angular velocity, shear force and bending moment but we shall use the particle velocity as the characterizing quantity. Assuming a solution in the form $v_y = \hat{v}_y e^{j(\omega t - k_B x)}$, the following expression for the wave number k_B is calculated as,

$$k_B = \frac{\omega}{c_B} = \sqrt{\omega} \cdot \sqrt[4]{\frac{m}{B}} \qquad (2.21)$$

The bending wave speed c_B is calculated by Equation **2.22** with h as plate thickness.

$$c_B = \sqrt{\omega} \cdot \sqrt[4]{\frac{B}{m}} = \sqrt{2\pi f} \cdot \sqrt[4]{\frac{Eh^2}{12\rho(1-v^2)}} \qquad (2.22)$$

We can see from the Equation **2.22** that the medium is dispersive for bending waves, which means that the phase speed is a function of frequency (i.e. frequency dependent). A broadband pulsed signal, therefore, change its shape during propagation (i.e. the high frequency wave components will outrun the components having a lower frequency). For a homogeneous plate having a thickness h we get the wave speed as,

$$c_B = \sqrt{\frac{\pi}{\sqrt{3}} c_L h f} \qquad (2.23)$$

Where, f is the frequency in Hz and c_L is the phase speed for the longitudinal waves in the medium, given in Equation **2.16**. We arrive at this expression by substituting for the quantities, $m = \rho h$ and $B = \frac{Eh^3}{12(1-v^2)}$. It is also noted that the expression given for the wave number and phase speed presuppose that the plate is thin, therefore, the wavelength of the bending wave should be larger than six time the plate thickness h. In other words, c_B should be less than $0.3c_L$.

2.3.5. Loss Factor for Bending Waves: (Internal Energy Losses in Materials)

In order to estimate the internal loss factor η of a plate we need to introduce complex bending wave stiffness defined as $B' = B(1 + j\eta)$. Here, η is the loss factor which is defined as the ratio of the mechanical energy E_d dissipated in a period of vibration to the reversible mechanical energy E_m and is defined as $\eta = \frac{E_d}{2\pi E_m}$. The loss

19

factor η may be determined by measuring either the quality factor Q, the bandwidth Δf for a given mode at resonance or the structural reverberation time T_S by following an excitation of the system at the given natural frequency f_0. The relations between these quantities are given as,

$$\eta = \frac{1}{Q} = \frac{\Delta f}{f_0} = \frac{2.2}{T_S f_0} \qquad (2.24)$$

The energy losses of a given element, however, always be caused by several mechanisms; such as, inner losses in the material (that usually very small of the order of 10^{-3}), energy as sound and leakage to connected structures or edge losses. The total loss factor, therefore, is expressed by a sum of all loss factors, as given by Equation **2.25**.

$$\eta_{tot} = \eta_{int} + \eta_{rad} + \eta_{edge} \qquad (2.25)$$

Most of the constructions composed of concrete, gypsum etc. and the loss factor due to the edge losses tends to dominate. This is, however, the important factor when it comes to sound transmission and its estimation and measurement in the field. According to ISO [6] the expression for the total loss factor can written as,

$$\eta_{tot} = \eta_{int} + \frac{\rho_0 c_0 \sigma}{\pi f m} + \frac{c_0}{\pi^2 S \sqrt{f f_c}} \sum_{k=1}^{4} l_k \alpha_k \qquad (2.26)$$

2.3.6. Critical frequency

The acoustic wave speed in air is independent of the frequency ($c_0 \approx 344\,\frac{m}{s}$ at $20°C$), whereas, the bending wave speed and the effective transverse wave speed are frequency dependent. This has consequence that there exists a frequency where the two wave types have the same speed and wavelength and therefore couple easily. The critical frequency f_c, defined as the frequency when $c_B = c_0 = 344\,\frac{m}{s}$, is given in Equation **2.22**. Where, K_c is the material constant and it is noted that for a given material, the critical frequency is inversely proportional to the plate thickness, h.

$$f_c = \frac{c_0^2}{2\pi}\sqrt{\frac{m}{B}} = \frac{c_0^2}{\pi h}\sqrt{\frac{3\rho(1-v^2)}{E}} = \frac{K_c}{h} \qquad (2.27)$$

2.4. Sound Radiation from Building Elements

The sound insulation of a building element or a complex construction, either for airborne sound or impacts depends on two factors: 1) the dynamic response to the actual excitation, being an acoustic field or a direct mechanical force or moment and 2) the efficiency as a sound radiator given the actual response pattern [20,22]. In this section, the sound radiation from plates is described for both forced and resonant vibrations, and the application of Rayleigh's method of radiation calculation is demonstrated in the case of radiation from forced bending waves. Forced vibrations are the part of the vibrations that are directly due to the surrounding sound field exciting the plate. In contrast, resonant vibrations are the free vibrations caused by reflections of the forced vibrations from the boundaries. We deal with the sound radiation from plane elements when given a bending wave velocity distribution. We discuss the definition of a quantity that is used to characterize the efficiency of a surface as a sound radiator known as, the radiation factor or radiation efficiency.

2.4.1. Radiation Factor (Radiation Efficiency)

A commonly used quantity to characterize the efficiency of a given vibrating surface, as a sound radiator is the radiation factor σ also called radiation efficiency or radiation ratio, defined in Equation **2.28**. Where, W_{rad} is the radiated power from the actual vibrating surface having the area S to the surrounding medium with characteristic impedance $\rho_0 c_0$. The quantity $\langle \tilde{u}^2 \rangle$ is the mean square velocity amplitude taken over the surface.

$$\sigma = \frac{W_{rad}}{\rho_0 c_0 S \langle \tilde{u}^2 \rangle} \tag{2.28}$$

The denominator in the expression is the power radiated from a partial area S of an infinitely large plane surface, all parts vibrating in phase with a velocity equal to this mean value, i.e. a plane wave radiation condition. Here we refer to the calculation of the radiated power to [20]. This expression shows the same expression when the radiating piston dimensions become much larger than the wavelength. The mean squared velocity in most of the literature is taken as average in the spatial domain, i.e. of the square RMS-value taken over all points on the surface. The argument given for this spatial averaging is that in a practical sense the velocity does not vary too much from point to point, making it sensible to represent the velocity as a mean value.

2.4.2.   Sound Radiation from an Infinite Large Plate

To calculate the sound pressure p in a point with coordinates (x, y) above the plate and the radiation factor σ, the bending wave velocity of a vibrating plate is taken as $u_B = \hat{u}e^{j(\omega t - k_B x)}$. Here, k_B is the wave number of the bending wave propagating in x direction. The sound pressure is given by Equation **2.29** and the bending wave is shown in Figure **2.2**.

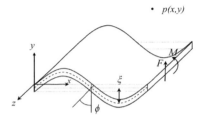

Figure 2.2: *Plane bending wave on an infinitely large plate. The plate lies in the xz-plane and the pressure is calculated at point (x, y)*

$$p(x, y) = \hat{p}e^{j(\omega t - k_x x - k_y y)} \tag{2.29}$$

k_x and k_y are the components of the wave number in the medium around the plate in air. The expression has then to be a solution of the ordinary wave equation, $\nabla^2 p - \frac{1}{c_0^2}\frac{\partial^2 p}{\partial t^2} = 0$. After using bending wave velocity in Equation **2.23**, the wave number k of a wave is calculates as,

$$k = \frac{\omega}{c_0} = \sqrt{k_x^2 + k_y^2} \tag{2.30}$$

A further condition is that the component v_y of the particle velocity must be equal to u_B at the surface of the plate $(y = 0)$. Since v_y is given by,

$$v_y = -\frac{1}{j\omega\rho_0}\frac{\partial p}{\partial y} = \frac{\hat{p}k_y}{\rho_0\omega}e^{j(\omega t - k_x x - k_y y)} \tag{2.31}$$

After setting $y = 0$, we get the following expression,

$$\hat{u}e^{-jk_Bx} = \frac{\hat{p}k_y}{\rho_0\omega}e^{-jk_xx} \qquad (2.32)$$

Hence

$$\hat{p} = \frac{\rho_0\omega}{k_y}\hat{u} \qquad (2.33)$$

and $k_x = k_B$. The sound pressure may now be expressed as,

$$p(x,y) = \frac{\rho_0 c_0 \hat{u}}{\sqrt{1 - \frac{k_B^2}{k^2}}} e^{j(\omega t - k_B)} e^{j\sqrt{k^2 - k_B^2}\,y} \qquad (2.34)$$

The Equation **2.34** shows that the important factor for the sound radiation is the ratio of the wave numbers in the plate and the surrounding medium. When $k_B > k$, i.e. the wavelength λ_B in plate is smaller than the wavelength λ in the air, the sound pressure will decrease exponentially with the distance y. On the other hand, if $k_B < k$ $(\lambda_B > \lambda)$, we have an ordinary propagating plane wave where the sound pressure increases with increasing ratio, $\frac{k}{k_B}$. This may be expressed by the angle ϕ of the radiated wave as,

$$\frac{k}{k_B} = \frac{1}{\sin\phi} \quad \text{or} \quad \left(\lambda_B = \frac{\lambda}{\sin\phi}\right) \quad \text{or} \quad \lambda = \lambda_B \sin\phi \qquad (2.35)$$

Therefore, the radiation factor is given by Equation **2.36**, by assuming $k_B < k$.

$$\sigma = \frac{1}{\sqrt{1 - \frac{k_B^2}{k^2}}} \qquad (2.36)$$

2.4.3. Sound Radiation from a Finite Plate

Below the critical frequency, the plate follows the mass movement induced by the incident sound pressure which is called "forced vibration" or "forced transmission". At low frequencies (less than the critical frequency) no radiation from bending waves can be expected for infinite plates, but this is not the case for finite real plate of finite size. In practice, we always have finite size plates of finite dimensions which is more complicated than the idealized example with the infinite plate. As illustrated in [20], in case of a finite plate we can assume that the vibration of the plate is determined by its natural modes, the radiation will depend on actual mode patterns which is determined by the modes taking part and their individual vibration amplitudes. Therefore, we cannot determine the radiation factor form the dimensions and material properties of the plate.

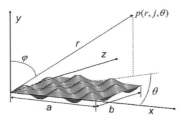

Figure 2.3: *Sound radiation from a finite size plate* [20]

With a stationary mechanical excitation, the structure will be forced into vibration by a more or less broad banded source which means that the vibration pattern is a combination of the natural modes having eigenfrequencies inside the actual frequency band being excited into resonance. In case of rectangular plate, we can assume that all modes having their natural frequency within the actual frequency band have the same velocity amplitude. It is quite useful to calculate the radiation factor for a single mode to see how critical the vibration pattern is concerning the radiated power. We, therefore, calculate the radiation factor for a simply supported plate set in an infinite baffle. With the assumption that the plate is vibrating in a simple harmonic way, the velocity is given by,

$$u_y(x,z) = \hat{u}\sin\left(\frac{n_x\pi x}{a}\right)\sin\left(\frac{n_z\pi z}{b}\right), \quad 0 \leq x \leq a, 0 \leq z \leq b \qquad (2.37)$$

Where n_x and n_z are the modal numbers in x and z axis, respectively, that is Figure **2.3**, where the plate vibrates in a (**5,4**) mode.

$$k_{n_x,n_z} = \left[\left(\frac{n_x\pi}{a}\right)^2 + \left(\frac{n_z\pi}{b}\right)^2\right]^{\frac{1}{2}} = \left[k_{n_x}^2 + k_{n_z}^2\right]^{\frac{1}{2}} \qquad (2.38)$$

The corresponding eigenfrequencies are given by Equation **2.39** with B and m are the bending stiffness per unit length and mass per unit area, respectively.

$$f(n_x, n_z) = \frac{\pi}{2} \sqrt{\frac{B}{m} \left[\left(\frac{n_x}{a} \right)^2 + \left(\frac{n_z}{b} \right)^2 \right]} \tag{2.39}$$

Based on the Rayleigh integral introduced in [23], the calculated sound pressure and the intensity in the far field from a plate where the velocity is given by Equation **2.37** and integrating the intensity over a hemisphere over the plate, we get the radiated power and thereby the radiation factor by using Equation **2.36**. Here, $\alpha = ka\sin\phi\cos\theta$ and $\beta = kb\sin\phi\cos\theta$

$$\sigma(n_x, n_z) = \frac{64k^2 ab}{\pi^6 n_x^2 n_z^2} \int_0^{\frac{\pi}{2}} \int_0^{\frac{\pi}{2}} \left\{ \frac{\frac{\cos \left[\frac{\alpha}{2} \right]}{\sin \left[\frac{\alpha}{2} \right]} \cdot \frac{\cos \left[\frac{\beta}{2} \right]}{\sin \left[\frac{\beta}{2} \right]}}{\left[\left(\frac{\alpha}{n_x \pi} \right)^2 - 1 \right] \cdot \left[\left(\frac{\beta}{n_z \pi} \right)^2 - 1 \right]} \right\}^2 \sin\phi \, d\phi \, d\theta \tag{2.40}$$

The frequency averaged radiation factor for a plate with dimensions a and b where $a < b$ and with f_c as critical frequency, can be calculated as given by Equation **2.41** is similar as given in ISO [6].

$$\sigma = \frac{2(a+b)c_0}{2\pi^2 \sqrt{f f_c} ab \sqrt{\frac{f_c}{f} - 1}} \left[\ln \left(\frac{\frac{f_c}{f} + 1}{\frac{f_c}{f} - 1} \right) + \frac{\frac{2f_c}{f}}{\frac{f_c^2}{f^2} - 1} \right] \quad \text{for} \quad f < f_c$$

$$\sigma = \sqrt{\frac{2\pi f}{c_0}} \sqrt{a} \left[0.5 - \frac{0.15a}{b} \right] \quad \text{for} \quad f \cong f_c \tag{2.41}$$

$$\sigma = \frac{1}{\sqrt{1 - \frac{f_c}{f}}} \quad \text{for} \quad f > f_c$$

However, there is no single formula to compute the radiation efficiency from the formula available in literature. It should be noted that the critical frequency should be much higher than the first eigenfrequencies of the plate. The formulas are only applied to a resonant multimode vibration of a plate. An example of radiation efficiency, given by [20] (in Figure **6.15**), is calculated by using Equation **2.41** where the author used a sheet of aluminium with ne length equals 2m and other varies from 0.2m to 5m.

In the frequency range below the critical frequency f_c the radiation factor is larger for the sound field excitation than for a mechanical excitation. In this case, the wave field in the plate is partly determined by the sound pressure distribution imposed by the sound field, which is a forced vibration field, and partly by the free waves originating from the edges of the finite plate as illustrated in [20]. The non-resonant part will be dominant when it comes to sound radiation which means that we can predict the sound transmission through a panel or wall by taking both the resonant and the non-resonant radiation into account. The radiation factor for forced vibrations by a sound field is necessarily dependent on panel dimensions and the actual wavelength, but also on the angle of sound incidence. In building acoustics, generally, the primarily interests are in the radiation factor for an incident diffuse field, however, in this thesis later on, we will also discuss and implement angle dependent radiation efficiency. For the diffuse sound field several alternative expressions exist in the literature. For example, Sewell [27] and Ljunggren [36] proposed the following expression for the force transmission, which is written as

$$\sigma_f = \frac{1}{2}\left[\ln\left(k\sqrt{S}\right) + 0.16 - F(\Lambda) + \frac{1}{4\pi k^2 S}\right] \tag{2.42}$$

$\Lambda = \frac{b}{a}$ ($\Lambda > 1$) and $F(\Lambda)$ are shape functions. Based on this equation, ISO [6] gives an approximate formula, where an upper limit of 2 is applied to the value of σ_f, i.e. $10\lg(\sigma_f)$ has a maximum value of 3 dB.

In the next chapter the basic sound propagation theory in air and structures will be combined in order to discuss a sound insulation model for prediction of the transfer functions between source and the receiver.

3

Airborne Sound Insulation Models

Sound is transmitted mostly through walls, ceilings and floors by setting the entire structure into vibration. When a sound wave is incident on the surface of a wall or any other surface which is separating two adjacent rooms, it is partly reflected back to the source domain and partly dissipated as heat within the material of the wall. Some sound energy, however, propagates to other connecting structures and some is partly transmitted into the receiving room domain. In this chapter, the quantities are introduced which characterize the airborne sound transmission for single monolithic plates. The quantities that are found in common building regulations and requirements for the sound insulation properties of building elements and constructions are presented. However, structure-borne sound transmission is not in the scope of this dissertation. The objective is to design airborne sound insulation models based on available prediction methods and research for an upgrade towards real-time auralization of different indoor and outdoor environments. The proposed (upgraded) sound insulation model presented in this chapter does not claim to be superior to the existing software tools. Therefore, it is discussed the fundamental concept of sound insulation prediction techniques from the filter design and auralization perspectives.

3.1. Airborne Transmission (Sound Reduction Index)

The transmission factor τ (or sound transmission coefficient) of a given surface (a wall element) is defined by means of sound power, i.e. the ratio of the transmitted power W_t from this element and the incident power W_i on its surface as given by Equation **3.1(a)**. However, the sound transmission coefficients are typically very small in numbers, therefore, it is more convenient to use the term "sound reduction index" denoted by R (which is also known as sound transmission loss) with the units in decibel (dB), and is defined in Equation **3.1(b)**.

$$\tau = \frac{W_t}{W_i} \tag{3.1-a}$$

$$R = 10 \lg \left(\frac{1}{\tau}\right) = 10 \lg \left(\frac{W_i}{W_t}\right) \tag{3.1-b}$$

Under diffuse-field conditions, as proposed by ISO [6], the sound field in the source room as well as in the receiving room is assumed to be diffuse. Which means that if $\tilde{p}_s$ is the sound pressure in the source room, the sound intensity at the surface of any wall element of the room is calculated by Equation **3.2**.

$$I_i = \frac{\tilde{p}_s^2}{4 \rho_0 c_0} \tag{3.2}$$

Therefore, the sound power transmitted through a building element having the surface area S is calculated by Equation **3.3**, with p_R as the sound pressure in the receiving room and A_R is the total absorption area of the receiving room.

$$W_t = \frac{\tilde{p}_R^2 \cdot A_R}{4 \rho_0 c_0} \tag{3.3}$$

The sound reduction index, denoted by R, is hence given by Equation **3.4**.

$$R = 10 \lg \left(\frac{1}{\tau}\right) = 20 \lg \left(\frac{\tilde{p}_s}{\tilde{p}_R}\right) + 10 \lg \left(\frac{S}{A_R}\right) = L_S - L_R + 10 \lg \left(\frac{S}{A_R}\right) \tag{3.4}$$

In Equation **3.4**, the difference $D = L_S - L_R$ is the difference in the mean sound pressure level in the sending and receiving room. Nevertheless, this expression is used in a standard laboratory procedure based on measurements of the sound pressure levels (e.g. ISO-140), however, the real situation is different than laboratory. Alternative methods are based on determination of the transmitted power to the receiving room by measuring the intensity. This expression is used when traditional methods fail because of flanking transmission. With this method, we can determine the mean transmitted intensity I_R over a surface S_R that completely encloses the actual element having an area S. Therefore, we get a new expression for sound reduction index R, given in Equation **3.5**.

$$R_I = 10 \lg \left(\frac{\tilde{p}_S^2}{4 \rho_0 c_0 I_R} \right) + 10 \lg \left(\frac{S}{S_R} \right) \cong L_{pS} - L_{IR} + 10 \lg \left(\frac{S}{S_R} \right) - 6 \, dB \qquad (3.5)$$

Usually, in buildings, there are many transmission paths available for the sound energy to travel from source to the receiving room. The sound energy, in addition to being directly transmitted through the main separating element (direct partition), may also be transmitted via flanking constructions, crack formations, out and in through windows, common ventilation ducts and cable ducts etc. As it is not possible to quantify all the possible transmission path, therefore, the sound reduction index in the receiving room, can be written as,

$$R' = D + 10 \lg \left(\frac{S}{A_R} \right) \qquad (3.6)$$

where D is sound pressure level difference between rooms and the quantity R' (sound reduction index) is now known as apparent sound reduction index of the partition. The sound pressure level difference referred to a given reverberation time T is denoted as the standardized level difference with T_0 set to 0.5 seconds for dwellings. It can be given by Equation **3.7**.

$$D_{nT} = D + 10 \lg \left(\frac{T}{T_0} \right) \qquad (3.7)$$

3.2. Direct Transmission

This section starts by looking at airborne sound insulation of a solid homogeneous isotropic plate as a base from there we discuss some of the building elements that are encountered in practice. In a complete building there are different transmission paths that determine the overall sound insulation performance such as direct transmission and flanking transmission. Flanking transmission is rather complex as compared to direct transmission. Therefore, both are discussed in separate sections. Direct sound transmission across a single building element occurs where the element is excited by an airborne sound source on one side. Figure **3.1** illustrates the concept of direct sound transmission. To predict airborne sound transmission, it is required to calculate the bending wave field induced by the excitation and thereafter find the resulting radiated power due to this field. However, the vibration pattern of the structure is more complex having two components that are 1) a forced-vibration field, which is, imparted to the wall due to the external sound field (also called the non-resonant field) and 2) a resonant field, which is a vibration field due to the natural modes excited by reflections from the boundaries [20]. The radiated sound power, generally, may now be expressed as,

$$W = \rho_0 c_0 S \{ \langle \tilde{u}_f^2 \rangle \, \sigma_f + \langle \tilde{u}_r^2 \rangle \, \sigma_r \} \tag{3.8}$$

The indices f and r in Equation **3.8** indicate the radiation factor (or efficiency) for "forced" and "resonant" transmissions respectively of building element with S as its surface area, and $\tilde{u}_f$ and $\tilde{u}_r$ are corresponding vibration velocities. We need to treat this case theoretical because of two types of vibration patterns on structures and due to the dependency of the angle of incidence of the sound field. Afterwards, we proceed to the calculation procedures covering the case of most commonly used airborne sound transmission by a diffuse field for an infinite and a finite plate.

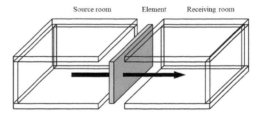

Figure 3.1: *An example of source room, receiver room and the separating wall element (main partition) for direct sound transmission* [15]

3.2.1. Direct Transmission: Infinite Plate

Most of the walls and floor constructions are far more complex than idealized forms. Yet an understanding of sound transmission through a simple plate is of fundamental importance as it is often used as a benchmark for comparison with more complex constructions. The main features of sound transmission can also be explained by considering a plate of finite thickness but infinite sizes. In practice, many problems in sound insulation design, prediction, and measurements revolve around the finite size of plates and their connections to other plates. We discuss different cases for calculation of sound reduction index in the following sections.

3.2.1.1. Direct Transmission Characterized by Mass Impedance

A thin wall or plate is better considered as a membrane (without tensional forces) or a collection of loosely connected point masses (i.e. a plastic curtain or something comparable). At this point, for simplicity, let us assume normal sound incidence on the plate. The resulting input impedance Z in this case as described in [20] is given by Equation **3.9**, which is a linear combination of the mass impedance of the plate and the characteristic impedance of the air behind the plate.

$$Z = \rho_0 c_0 + j\omega m = Z_0 + j\omega m \qquad (3.9)$$

The plate represents a boundary surface and gives an absorption factor α that is calculated using the following Equation **3.10**.

$$\alpha = \frac{4Re\left\{\frac{Z}{Z_0}\right\}}{\left|\frac{Z}{Z_0}\right|^2 + 2Re\left\{\frac{Z}{Z_0}\right\} + 1} \qquad (3.10)$$

From characterization of the plate by its mass impedance and having no internal energy losses, the transmission factor τ of the plate must be equal to the absorption factor α. Using Equation **3.9** in Equation **3.10** we get,

$$\tau = \frac{1}{1 + \left(\frac{\omega m}{2\rho_0 c}\right)^2} \qquad (3.11)$$

and corresponding sound reduction index R_0 in Equation **3.12**.

$$R_0 = 10\lg\left(\frac{1}{\tau}\right) = 10\lg\left[1 + \left(\frac{\omega m}{2\rho_0 c}\right)^2\right] \cong 20\lg\left(\frac{\pi f m}{\rho_0 c}\right) \qquad (3.12)$$

This is known as the so-called "mass law" in its simplest form which means that the sound reduction index increases by 6 dB by each doubling of frequency and/or mass per unit area. The approximation given by last expression in Equation **3.12**, however, presuppose that the mass impedance is much larger than the characteristic impedance of air. This condition is normally fulfilled for panels used in buildings. Inserting the characteristic impedance of air at 20°C we get the simplified version of R_0 given as

$$R_0 = 20 \lg(mf) - 42.5 \; dB \tag{3.13}$$

3.2.1.2. Bending Wave Field: Characterized by Wall Impedance

Before looking at the bending waves on an infinite plate, it is useful to consider sound radiation from a plane bending wave propagating without damping. At the critical frequency equal phase velocities correspond to $\lambda_B = \lambda$. The critical frequency describes the lowest frequency at which coincidence occurs. Consider a bending wave of stiffness per unit length as B and mass per unit area as m to solve the wave equation given in Equation **3.14**, where the sound pressure of the incoming wave is the driving force.

$$B\nabla^2\nabla^2\xi + m\frac{\partial^2\xi}{\partial^2 t} = p(x,y,z,t) \tag{3.14}$$

The quantity ξ is the particle displacement which is known as the deflection of the plate surface. Assuming a harmonic time function $e^{j\omega t}$ and using the velocity u as a variable we get,

$$\nabla^2\nabla^2 u - k_B^4 u = \frac{j\omega}{m}p(x,z) \tag{3.15}$$

Let us assume that a plane wave hits the surface of the plate at an angle θ, so that the condition $\lambda_B = \lambda$ is achieved, the sound pressure $p(x,z)$ can be casted in the form of Equation **3.16**.

$$p(x,z) = \hat{p}(k_x,k_z)e^{jk_x x}e^{jk_z z} \tag{3.16}$$

Inserting Equation **3.16** into Equation **3.15**, we obtain the following relation between the amplitudes of the pressure and the velocity given in Equation **3.17**, which gives an important relationship between the acoustic wave number and the bending wave number.

$$\hat{u}(k_x) = \frac{j\omega \cdot \hat{p}(k_x)}{B(k_x^4 - k_B^4)} \tag{3.17}$$

This becomes more evident when we calculate the velocity. The incident, reflected and transmitted sound pressure are represented by Equation **3.18**(a-c) respectively with the total pressure on the plate in Equation **3.18**(d).

$$p_i = \hat{p}_i e^{-jky\cos\theta} e^{-jkx\sin\theta} \tag{3.18-a}$$

$$p_r = \hat{p}_r e^{jky\cos\theta} e^{-jkx\sin\theta} \tag{3.18-b}$$

$$p_t = \hat{p}_t e^{-jky\cos\theta} e^{-jkx\sin\theta} \tag{3.18-c}$$

$$p(x,z) = (\hat{p}_i + \hat{p}_r - \hat{p}_t) e^{-jkx\sin\theta} \tag{3.18-d}$$

With $k_x = k\sin\theta$, we get the equation giving the relationship between the driving sound pressure and the resulting velocity is given as,

$$\hat{u} = \frac{j\omega(\hat{p}_i + \hat{p}_r - \hat{p}_t)}{B(k^4\sin^4\theta - k_B^4)} \tag{3.19}$$

The ratio of the driving pressure to the velocity is generally known as wall impedance. This quantity, $Z(\theta)$ is given by,

$$Z(\theta) = \frac{(\hat{p}_i + \hat{p}_r - \hat{p}_t)}{\hat{u}} = \frac{B}{j\omega}(k^4\sin^4\theta - k_B^4) \tag{3.20}$$

The impedance $Z(\theta)$ becomes zero under the condition $k > k_B$ at an incident angle θ, making the velocity 'infinitely' large. Which means that the plate will not present any obstacle for the sound wave. Introducing the critical frequency f_c and the energy losses by way of a complex bending stiffness $B(1 + j\eta)$, we can write Equation **3.20** as follows,

$$Z(\theta) = j\omega m \left[1 - \left(\frac{f}{f_c}\right)^2 . (1 + j\eta) \sin^4 \theta \right] \qquad (3.21)$$

The total loss factor η of a single wall is defined as the sum of the internal damping loss factor of panel, the damping loss factor due to the transmission of vibrational energy from the panel to its surrounding elements at its edges and twice its single sided radiation loss factor, which is given b Equation **3.22**.

$$\eta = \eta_{int} + \eta_{edge} + 2\eta_{rad} \qquad (3.22)$$

The single radiation loss factor is corresponding to the single sided radiation factor by Equation **3.23**.

$$\eta_{rad} = \frac{\sigma(\theta)\rho_o c}{m\omega} \qquad (3.23)$$

3.2.1.3. Direct Transmission (Angle Dependent)

The sound reduction index R (transmission factor τ) is calculated from the ratio of the sound pressure amplitudes in the transmitted and incident waves written as $\tau = \frac{W_t}{W_i}$. Using Equation **3.19**, the velocity can be rewritten as,

$$u = \hat{u} e^{-jkx \sin \theta} = \frac{(\hat{p}_i + \hat{p}_r - \hat{p}_t)}{Z_w} . e^{-jkx \sin \theta} \qquad (3.24)$$

The normal component of the acoustic particle velocity v on both sides of the plate must be equal to the plate velocity u. Hence, the following relationship must apply $\hat{v}_i + \hat{v}_r = \hat{u} = \hat{v}_t$. The relationship between these velocity amplitudes and the corresponding pressure amplitudes is easily found by applying the force equation (Euler equation).

$$v_{y=0} = -\frac{1}{j\omega\rho_0} \left(\frac{\partial p}{\partial y}\right)_{y=0} \qquad (3.25)$$

From Equations **3.25**, we can write the components of the incident, reflected and transmitted velocities by, $\hat{v}_i = \frac{\hat{p}_i}{z_0} \cos \theta$, $\hat{v}_r = -\frac{\hat{p}_r}{z_0} \cos \theta$ and $\hat{v}_t = \frac{\hat{p}_t}{z_0} \cos \theta$, respectively. Using these velocity components with Equation **3.24** and the velocities conditions, the relationship between pressure amplitudes can be found as,

$$\hat{u}(\theta) = \frac{2\hat{p}_i}{Z(\theta) + \dfrac{2\rho_o c}{\cos\theta}} = \frac{\hat{p}_t}{\rho_o c}\cos\theta \tag{3.26}$$

The transmission factor (reduction index) will then be given by with $\sigma(\theta) = \frac{1}{\cos\theta}$

$$\tau(\theta) = \frac{1}{\left|1 + \dfrac{Z(\theta)\cos\theta}{2\rho_o c}\right|^2} = \frac{1}{\left|1 + \dfrac{Z(\theta)}{2\rho_o c \cdot \sigma(\theta)}\right|^2} \tag{3.27}$$

$$R(\theta) = 10\lg\left(\frac{1}{\tau}\right) = \left|1 + \frac{Z(\theta)\cos\theta}{2\rho_o c}\right|^2 \tag{3.28}$$

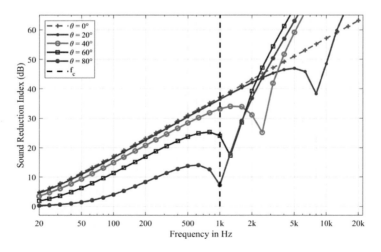

Figure 3.2: *Sound reduction index of an infinitely large plate with the angle of incidence as parameter (for resonance transmission)*

The wall impedance $Z(\theta)$ is given by Equation **3.21**. An example of angle dependent sound reduction, using these equations, is shown in Figure **3.2** for a plate of surface mass $m = 10 \frac{kg}{m^2}$, the critical frequency $f_c = 1000$ Hz, the internal loss factor $\eta = 0.1$ at several given incident angles. For an infinite panel, the single sided radiation efficiency is taken as $\sigma(\theta) = \frac{1}{\cos(\theta)}$. Inserting Equation **3.21** and $\sigma(\theta)$ into Equation **3.27** produces Cremer's [**31**] sound transmission coefficient $\tau(\theta)$ of a single leaf panel as a function of angle of incidence θ given in Equation **3.29**.

$$\tau(\theta) = \frac{1}{\left(1 + \frac{\pi f m \eta \cdot f^2 \sin^4 \theta}{\rho_0 c f_c \cdot \sigma(\theta)}\right)^2 + \left(\frac{\pi f m}{\rho_0 c \cdot \sigma(\theta)}\right)^2 \left(1 - \left(\frac{f}{f_c}\right)^2 \sin^4 \theta\right)^2} \tag{3.29}$$

Davy in [**25**] approximated the expression for angle dependent transmission coefficient in Equation **3.29**, for a given frequency which is greater than or equal to the critical frequency. The maximum value of transmission coefficient Equation **3.29** occurs at the coincidence angle θ_c, when $\sin^2 \theta_c = \frac{f_c}{f}$. For the values of θ, which are close to θ_c, Equation **3.29** is approximated by setting most of the values of θ equal to θ_c, and results as,

$$\tau(\theta) = \frac{1}{\left(1 + \frac{\pi f m \eta}{\rho_0 c \cdot \sigma(\theta_c)}\right)^2 + \left(\frac{2\pi f_c m}{\rho_0 c \cdot \sigma(\theta_c)}\right)^2 \left(\frac{f_c}{f} - \sin^2 \theta\right)^2} \tag{3.30}$$

Using $r = \frac{f}{f_c}$, $a = \frac{\omega m}{2\rho_0 c} = \frac{\pi f m}{\rho_0 c}$ and substituting $x = \cos^2 \theta$ then $\sin^2 \theta = 1 - x^2$ Equation **3.30** gives,

$$\tau(\theta) = \frac{1}{\left(1 + \frac{a\eta}{\sigma(\theta_c)}\right)^2 + \left(\frac{2ar}{\sigma(\theta_c)}\right)^2 \left(x + \frac{1}{r} - 1\right)^2} \tag{3.31}$$

3.2.1.4.   Direct Transmission (Diffuse Field)

The incident sound on a partition, separating two adjacent rooms, is normally considered arriving from many directions at the same time due to the reflections from the other walls elements, which is the so-called diffuse field assumption. We can use Equations **3.31**, by making weightings according to the given distribution of incident angles and sum up the contributions. In practice, however, the actual distribution is rarely known, therefore, an ideal diffuse incident sound field may be assumed for calculations. Which mean, assuming that the sound incidence is evenly distributed over all angles and with random phases. Using integral over angle dependent transmission factor $\tau(\theta)$ and using $\sigma(\theta) = \frac{1}{\cos\theta}$, we get

$$\tau = 2\int_0^{\frac{\pi}{2}} \tau(\theta)\sin\theta\cos\theta\,d\theta = 2\int_0^{\frac{\pi}{2}} \frac{\tau(\theta)}{\sigma(\theta)}\sin\theta\,d\theta \tag{3.32}$$

Putting Equations **3.31** into Equations **3.32** gives the diffuse field sound transmission coefficient as

$$\tau = \int_0^1 \frac{dx}{\left(1 + \frac{a\eta}{\sigma(\theta_c)}\right)^2 + \left(\frac{2ar}{\sigma(\theta_c)}\right)^2 \left(x + \frac{1}{r} - 1\right)^2} \tag{3.33}$$

Now put $y = x + \frac{1}{r} - 1$ then $dy = dx$ and Equation **3.33** becomes

$$\tau = \int_{\frac{1}{r}-1}^{\frac{1}{r}} \frac{dy}{\left(1 + \frac{a\eta}{\sigma(\theta_c)}\right)^2 + \left(\frac{2ar}{\sigma(\theta_c)}\right)^2 y^2} \tag{3.34}$$

Davy in [25] used approximation (which are different than [31]) in Equation **3.34** and solved the integral from Ryzhik et. al. [30], whereas, Cremer [31] approximated the integral in Equation **3.34** by extended the limits of integration to $-\infty$ to $+\infty$. Hence, the final expression for the transmission factor derived in [25] is given in Equation **3.35**.

$$\tau = \frac{\sigma^2(\theta_c)}{2ar(\sigma(\theta_c) + a\eta)}\left\{\tan^{-1}\left[\frac{2a}{\sigma(\theta_c) + a\eta}\right] - \tan^{-1}\left[\frac{2a(1-r)}{\sigma(\theta_c) + a\eta}\right]\right\} \tag{3.35}$$

In [31], the approximation is derived on the bases that the integral in Equation **3.34** becomes maximum when $y = 0$. If $\frac{a\eta}{\sigma(\theta)} \gg 1$, which is usually the case, the integrand is half its maximum value when $|y| = \frac{\eta}{2r}$. Since η is usually very much less than 1 and r is greater than or equal to 1 (if the frequency is greater than or equal to the critical frequency) the values of y where the integrand is signi cantly different from zero usually lie well inside the integral limits from $\frac{1}{r} - 1$ to $\frac{1}{r}$. Because of this, Cremer [31] approximated the integral in Equation **3.34** by extending the limits of integration from $-\infty$ to $+\infty$ and gives the transmission coefficient as,

$$\tau \approx \int_{-\infty}^{\infty} \frac{dy}{\left(1 + \frac{a\eta}{\sigma(\theta_c)}\right)^2 + \left(\frac{2ar}{\sigma(\theta_c)}\right)^2 y^2} \tag{3.36}$$

With this approximations (Ref: [31] for details), Equation **3.35** becomes

$$\tau \approx \frac{\pi \sigma^2(\theta_c)}{2ar(\sigma(\theta_c) + a\eta)} = \frac{\pi \left(\frac{\sigma(\theta_c)}{a}\right)^2}{2r \left(\frac{\sigma(\theta_c)}{a} + \eta\right)} \tag{3.37}$$

Cremer in [31] also assumed a usual case of $\frac{a\eta}{\sigma} \gg 1$ and gives the transmission coefficient equals to $\tau \approx \frac{1}{a^2} \frac{\pi}{2\eta} \frac{\sigma^2(\theta_c)}{r}$. As in [31] an infinite panel is assumed above the critical frequency f_c, hence, radiation efficiency of the free bending waves is also used above f_c for such an infinite panel. This is because the wavelength of the forced waves at coincidence is equal to the free bending wavelength. Therefore, radiation efficiency in this case is $\sigma(\theta_c) = \frac{1}{\cos(\theta_c)}$, which now given the approximated diffuse transmission coefficient as $\tau \approx \frac{1}{a^2} \frac{\pi}{2\eta} \frac{1}{r-1}$. (Furthermore, when $r > 2$, $\tau \approx \frac{1}{a^2} \frac{\pi}{2\eta} \frac{1}{r}$). This approximation ($\frac{a\eta}{\sigma} \gg 1$) is equivalent to the assumption that the radiation efficiency of a panel above the critical frequency is equal to unity. This dissertation will use Equation **3.35** for diffuse field transmission, however, Cremer's approximated diffuse transmission is mentioned in this section is rather for comparison with the Davy's [25] approximations.

3.2.2.   Direct Transmission: Finite Plate

For infinite planar structures, the wave approach is generally used for sound transmission at high frequencies. It calculates the sound reduction index of an infinite plane structure excited by a single or random incident plane wave. However, the calculated results are often quite different from the experimental curves as mentioned in [25]. The most frequently used way to reduce differences between experimental results and the wave approach, has been to limit the maximum incidence angle of the acoustic field incident on the infinite structure. This of course strongly influences the results, however, the general behaviour is not much improved at low frequencies and the slope of the transmission index remains unchanged [32,33]. Fahy [23] stated two main factors that cause the sound performance of a real bounded panel in a rigid baffle to differ significantly from the theoretical performance of an unbounded panel which are; the existence of structural modes and associated resonance frequencies and the diffraction by the aperture in the baffle that contains the panel. An additional difference between calculated and measured results comes from the influence of room boundaries on transmission through partitions that separate two rooms. Some authors developed expressions for the radiation efficiency of panels including the diffraction effect for finite size single wall at low frequency. Sato [34] and Rindel [35] developed an integral expression of the radiation efficiency in the special case of a rectangular panel in a baffle forced by a plane sound wave at oblique incidence. In order to obtain the transmission index for random incidence excitation Rindel [35] used mean radiation efficiency applied to the mass driven approximation of the transmission loss. Ljunggren [36] also used the same approach to obtain the radiation efficiency of a one-dimensional structure. On the other hand, several other authors developed expressions for the radiation efficiency of finite size panels using modal approach. Among them is Sewel [37] and recently Leppington [38,39], who have now given classical and well-validated results. Villot [24] proposed a technique based on a spatial windowing of plane waves in order to take into account the finite size of a plane structure in sound radiation calculation. In the following sections we discuss different approaches to predict sound transmission from a finite panel. These techniques calculate radiation efficiencies for a finite panel for both forced and resonant transmission and hence predict sound transmission coefficient.

3.2.2.1.   Davy's Theory

In section **3.2.1**, we discussed that Davy's [25] used Cremer's sound insulation prediction model for infinite panel and extended it for angle dependent sound transmission for infinite plate as a starting point but introduced the radiation efficiency for a finite wall instead of using $\frac{1}{\cos\theta}$ (Cremer's formula). We will briefly discuss the derivation of the radiation efficiency of Davy's Theory [25] for frequencies above and below the critical frequency

3.2.2.1.1.   Above the Critical Frequency

To drive the radiation efficiency, Davy at first calculated cosine of the coincidence angle. As Equation **3.35** is used as a correction term below the critical frequency, the cosine of the coincidence angle is set to zero for the frequencies below the critical frequencies. This gives,

$$g = \begin{cases} \cos\theta_c = \sqrt{1 - \dfrac{\omega_c}{\omega}} = \sqrt{1 - \dfrac{f_c}{f}} & \text{if } \omega \geq \omega_c \\ 0 & \text{if } \omega < \omega_c \end{cases} \tag{3.38}$$

The final derivation of the radiation efficiency given by Davy [25] is as follows.

$$\sigma(\theta_c) = \begin{cases} \dfrac{1}{\sqrt[n]{g^n + q^n}} & \text{if } 1 \geq g \geq p \\ \dfrac{1}{\sqrt[n]{(h - \alpha g)^n + q^n}} & \text{if } p > g \geq 0 \end{cases} \tag{3.39}$$

In Equation **3.39**, $q = \frac{2\pi}{k^2 S}$, with $k = \frac{2\pi f}{c}$ and S is the area of the finite panel, and $\alpha = \frac{h}{p} - 1$, where h and p are given in Equation **3.40** and Equation **3.41**.

$$p = \begin{cases} w\sqrt{\dfrac{\pi}{2ka}} & \text{if } w\sqrt{\dfrac{\pi}{2ka}} \leq 1 \\ 1 & \text{if } w\sqrt{\dfrac{\pi}{2ka}} > 1 \end{cases} \tag{3.40}$$

$$h = \dfrac{1}{\dfrac{2}{3}\sqrt{\dfrac{2ka}{\pi}} - \beta} \tag{3.41}$$

Here, $a = \frac{2S}{U}$ with U as the perimeter of the finite panel. The empirical constants in Equation **3.39**, Equation **3.40**, and Equation **3.41** are taken as $n = 2$, $w = 1.3$ and $\beta = 0.124$ to optimise the radiation efficiency results close to those calculated by Sato [34] with numerical approach.

3.2.2.1.2.   Below the Critical Frequency

Below the critical frequency, the sound transmission is calculated using the averaged diffuse field single sided radiation efficiency approach. Bending stiffness can be ignored by setting $\frac{f}{f_c} = 0$ in Equation **3.21**, which becomes $Z(\theta) = j\omega m$. Using this $Z(\theta)$ in Equation **3.27** and assuming that $\frac{Z(\theta)}{2\rho_0 c \cdot \sigma(\theta)} \gg 1$, we get

$$\tau(\theta) = \left(\dfrac{2\rho_0 c \cdot \sigma(\theta)}{m\omega}\right)^2 = \left(\dfrac{\sigma(\theta)}{a}\right)^2 \tag{3.42}$$

For the diffuse transmission put Equation **3.42** into Equation **3.32**, we get,

40

$$\tau = \frac{2}{a^2} \int_0^{\frac{\pi}{2}} \sigma(\theta) \sin\theta \, d\theta = \frac{2\langle\sigma\rangle}{a^2} \tag{3.43}$$

In Equation **3.43**, $\langle\sigma\rangle = \int_0^{\frac{\pi}{2}} \sigma(\theta) \sin\theta \, d\theta$. This integral is solved by substituting Equation **3.39** into Equation **3.43**, which finally gives the radiation efficiency below the critical frequency as,

$$\langle\sigma\rangle = \ln\left(\frac{1 + \sqrt{1 + q^2}}{p + \sqrt{p^2 + q^2}}\right) + \frac{1}{\alpha}\ln\left(\frac{h + \sqrt{h^2 + q^2}}{p + \sqrt{p^2 + q^2}}\right) \tag{3.44}$$

As discussed earlier, the bending stiffness was ignored by using $Z(\theta) = jm\omega$ in Equation **3.42**. To include the effects of the bending stiffness, the sound transmission below the critical frequency is calculated as the sum of Equation **3.43** and Equation **3.35** which is the new approach proposed by Davy in [**25**].

3.2.2.2. Spatial Windowing Technique

The principle of spatial windowing technique is to compute the radiated power where only a small area S (with length L_x, width L_y and thickness L_z) of an infinite structure contributes to the sound radiations. The application of this technique can be realized in the situation where separating element (i.e. partition) between the adjacent rooms is large. Assuming a propagating structural wave of wave number k_p, the velocity field in the wave number domain is defined by taking the spatial Fourier transform giving the following expression,

$$\hat{v}(k_x, k_y) = \hat{v}. L_z L_y \frac{\sin\left((k_x - k_p\cos\psi)\frac{L_x}{2}\right). \sin\left((k_y - k_p\sin\psi)\frac{L_y}{2}\right)}{\left((k_x - k_p\cos\psi)\frac{L_x}{2}\right). \left((k_y - k_p\sin\psi)\frac{L_y}{2}\right)} \tag{3.45}$$

Where, ψ is the azimuth angle for k_p. This result is used to calculate the radiated power (with reference to Fahy [**23**]) and is given in Equation **3.46**, With $k_x = k_r\cos(\phi)$ and $k_y = k_r\sin(\phi)$.

$$W(k_p, \psi) = \frac{\rho_0 c_0}{8\pi^2} \int_0^{k_0} \int_0^{2\pi} \frac{|\hat{v}(k_x, k_y)|^2}{\sqrt{k_0^2 - k_r^2}} k_0 k_r d\phi \, dk_r \tag{3.46}$$

The radiation factor is then calculated by double integral, which may be written as,

$$\sigma(k_p,\psi) = \frac{S}{\pi^2}\int_0^{k_0}\int_0^{2\pi}\frac{\sin^2\left((k_r\cos\phi - k_p\cos\psi)L_x\right).\sin^2\left((k_r\sin\phi - k_p\sin\psi)L_y\right)}{[(k_r\cos\phi - k_p\cos\psi)L_x]^2[(k_r\sin\phi - k_p\sin\psi)L_y]^2}$$
$$\times \frac{k_0 k_r}{\sqrt{k_0^2 - k_r^2}}\, d\phi dk_r \tag{3.47}$$

Villot [24] found that the dependence of the radiation efficiency on the angle ψ is slight. Therefore, in order to condense the results and present the variation of the radiation efficiency as a function frequency and angle of incidence θ the radiation efficiency is averaged over ψ, given in Equation **3.48**.

$$\langle\sigma(k_p)\rangle_\psi$$
$$= \frac{S}{2\pi^3}\int_0^{2\pi}\int_0^{k_0}\int_0^{2\pi}\frac{\sin^2\left((k_r\cos\phi - k_p\cos\psi)L_x\right).\sin^2\left((k_r\sin\phi - k_p\sin\psi)L_y\right)}{[(k_r\cos\phi - k_p\cos\psi)L_x]^2[(k_r\sin\phi - k_p\sin\psi)L_y]^2}$$
$$\times \frac{k_0 k_r}{\sqrt{k_0^2 - k_r^2}}\, d\phi dk_r d\psi \tag{3.48}$$

Furthermore, Villot [24] observed that for $k_0 L < 4$, the radiation efficiency of the finite structure (e.g. $1.4 \times 1.1\, m$) using spatial windowing filtering is lower than that of the infinite system for any angle of incidence and does not significantly vary with the angle of incidence θ, whereas, for $k_0 L > 4$, radiation efficiency of the finite structure increases with the angle of incidence θ, following the radiation efficiency of the infinite system up to a certain incidence angle and does not significantly vary afterwards. For an acoustic wave the wave number k_p is related to the incident angle θ by $k_p = k_0 \sin\theta$. Therefore, the transmission coefficient given in Equation **3.31** can now be determined for a finite size of wall element of any small component of it by inserting $\sigma(\theta)$ for frequencies greater than critical frequency. Below the critical frequency, the sound transmission coefficient is calculated using the average diffuse field single sided radiation efficiency approach given in Equation **3.44**, while the bending stiffness is ignored. To include the effects of the bending stiffness, the sound transmission below the critical frequency is calculated as the sum of Equation **3.43** and Equation **3.35**. The idea is to apply spatial windowing to radiation process and find its effect on incident pressure field by correcting transmission factor of infinite structure to obtain corresponding transmission factor of the finite structure. Thus using Equation **3.48** into Equation **3.31**, we get the final transmission coefficient for a finite plate. Another approach is proposed by Vigran [**29**] to obtain the transmission coefficient $\tau_s(\theta)$ for finite plate by calculating transmission coefficient $\tau(\theta)$ for infinite plate and then use Equation **3.49** to get finite $\tau_s(\theta)$ (here subscript s denotes a small patch or segment of the large wall). Here the phase information of the radiations is missing which we discuss in Chapter **4** in sound insulation filters section.

$$\tau_s(\theta) = \tau(\theta)(\sigma(k_0 \sin \theta) \cos \theta) \qquad (3.49)$$

3.2.2.3.   ISO Standard Approach

For common homogeneous building elements, ISO standard [**6**] provides a comprehensive formulation (Equation **3.50**) to calculate laboratory sound reduction index $R = -10 \lg \tau$ and the sound transmission factor for a finite size wall element under diffuse field conditions. The standard suggest that below the critical frequency f_c, the contribution of the forced transmission can be neglected for the flanking paths in Equation **3.50**. The total loss factor term used in this equation is influence by the laboratory condition and therefore, should be taken into account. The Equation **3.50** represents a finite size wall with dimensions given by the quantities a and b, whereas, η_{tot} is the total loss factor. The terms σ and σ_f are the radiation factors for resonant and non-resonant (i.e. forced) part of the sound transmission respectively. The radiation factor for forced transmission in Equation 3.50 is calculated by $\sigma_f = 0.5(\ln k_o \sqrt{l_1 l_2} - \Lambda); \sigma_f \le 2$, which is based on [**27**] with the assumption that l_1 should be greater than l_2. Here $k_o = \frac{2\pi f}{c_o}$ is wave number, l_1 and l_2 are length and width of panel and $\Lambda = -0.964 - \left(0.5 + \frac{l_2}{\pi l_1}\right) \ln \frac{l_2}{l_1} + \frac{5 l_2}{2 \pi l_1} - \frac{1}{4 \pi l_1 l_2 k_o^2}$.

$$\tau = \left(\frac{\rho_o c}{\pi f m}\right)^2 \left[2\sigma_f + \frac{(a+b)^2}{a^2+b^2} \sqrt{\frac{f_c}{f}} \cdot \frac{\sigma^2}{\eta_{tot}}\right] \qquad \text{for} \quad f < f_c$$

$$\tau = \left(\frac{\rho_o c}{\pi f m}\right)^2 \left[\frac{\pi \sigma^2}{2 \eta_{tot}}\right] \qquad \qquad \text{for} \quad f = f_c \qquad (3.50)$$

$$\tau = \left(\frac{\rho_o c}{\pi f m}\right)^2 \left[\frac{\pi f_c \sigma^2}{2 f \eta_{tot}}\right] \qquad \qquad \text{for} \quad f > f_c$$

The derivation of the radiation factor for the free waves (σ) is given in ISO 12354-1 (Annex B) [6], however, these derivations of free wave transmission factor are valid for a plate surrounded by an infinite baffle in the same plate for which the radiation factor is calculated. In real buildings, the walls and floors are usually surrounded by orthogonal elements which might increase the radiation efficiency below the critical frequency by a factor of 2 or 4 for edge modes or corner modes respectively. An alternate to this problem is discussed in the previous sections (**3.2.2.1**) of this chapter and in the thesis the proposed extended sound insulation model is based on calculation of radiation efficiency on these alternate methods. Apart from proposed alternate for radiation efficiencies, another alternate to ISO [6] is to neglect the contribution from the resonant transmission but to include slightly a simplified area effect. Fahy [**23**] suggested that for the frequency range $f < f_c$ the expression for sound reduction index, given in Equation **3.51**, may be used.

$$R_f = R_0 - 10\lg\left[\ln\left(\frac{2\pi f}{c_0}\sqrt{ab}\right)\right] + 20\lg\left[1 - \left(\frac{f}{f_c}\right)^2\right] \tag{3.51}$$

This expression represents the forced transmission and by inserting for R_0, we get the approximate solution given Equation **3.52**.

$$R_f \cong 20\lg(mf) - 10\lg\left[\ln\left(\frac{2\pi f}{c_0}\sqrt{ab}\right)\right] + 20\lg\left[1 - \left(\frac{f}{f_c}\right)^2\right] - 42\text{ dB} \tag{3.52}$$

Further approximation can be made by setting $\sigma \approx 1$, in the frequency range above the critical frequency, and by using the last expression in Equation **3.53**, to get the following expression for sound reduction index.

$$R = 20\lg(mf) + 10\lg\left[2\eta_{tot}\frac{f}{f_c}\right] - 47\text{ dB} \tag{3.53}$$

3.3. Flanking Transmission

In previous section we discussed with some exceptions the sound transmission through a specific building element. Sound reduction index is ideally an element specification, however, the boundary conditions of an element may have considerable influence on the results. The types and properties of the connections to adjoining constructions are important factors when specifying the transmission properties of a given element. In this section we deal with the interplay of building elements with the objective of predicting airborne sound transmission in real buildings, in which there are involved, normally, a number of transmission paths between source and the receiver. We look for models enabling us to predict the acoustic performance of the buildings based on the acoustic performance of each element which make up the complete structures.

3.3.1.   Apparent Sound Reduction Index

The prediction of the apparent or effective sound reduction index (R') between the adjoining rooms in dwellings, either airborne sound or structure borne sound, depends on the selection or presupposing of a model that includes all types of transmission paths (direct or indirect paths). The commonly used models are generally confined to neighbouring rooms (two rooms separated by wall or floor elements) to determine the flanking transmissions. It is also assumed in many sound insulation models that the sound transmission between adjacent rooms involves different transmission paths which are independent and that all the wave fields are diffuse. In this section we are primarily treating the airborne sound insulation due to the fact that it normally represents greater problems for prediction than the impact sound insulation. In addition, the data used when calculating the apparent sound reduction index, e.g. vibration reduction index at junctions, may directly be applied to impact sound problems. For each transmission path of airborne sound, we allocate a transmission factor. Referring these factors to the partition, we may express the apparent sound reduction index and corresponding transmission factor by Equation **3.54** and Equation **3.55** respectively. Here, the terms d and f refer to direct sound energy radiations and flanking sound energy radiations respectively, whereas the terms e and s refers to portal energy radiations (i.e. doors, windows etc.) and indirect airborne energy radiations respectively.

$$R' = 10 \lg \left(\frac{1}{\tau'}\right) \tag{3.54}$$

$$\tau' = \tau_d + \tau_f + \sum_{e=1}^{m} \tau_e + \sum_{s=1}^{k} \tau_s \tag{3.55}$$

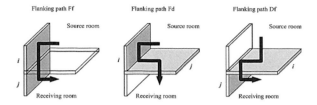

Figure 3.3: *Sound transmission paths in two adjacent rooms including flanking paths up to first-order junctions* [15]

To complete the picture, two terms are added representing the sum of other direct or indirect transmission. Figure **3.3** gives an indication on different transmission paths that are taken into account. It should be noted, however, that the sketch only indicates what we may denote first order flanking paths, i.e. the paths that involve one element in the sending room, one junction or connection and one element in the receiving room. Here, Dd is Direct-Direct path, Df is Direct-Flanking paths, Fd is Flanking-Direct paths and Ff is Flanking-Flanking paths.

$$\tau_d = \tau_{Dd} + \sum_{d=1}^{n} \tau_{Fd} \tag{3.56}$$

$$\tau_f = \sum_{f=1}^{n} \tau_{Df} + \sum_{f=1}^{n} \tau_{Ff} \tag{3.57}$$

3.3.2.   Flanking Sound Reduction Index

Considering that the flanking paths of the type Df are contributing in making use of the flanking reduction index which includes a direct transmission path through the main partition (i.e. the wall or floor which separates the adjoining rooms) with a surface area of S_D and sound reduction index R_d. Then, the apparent sound reduction index can be written in the form of Equation **3.58**.

$$R' = -10 \lg \left[10^{-\frac{R_d}{10}} + \sum 10^{-\frac{R_f}{10}} \right] \tag{3.58}$$

The flanking sound reduction index R_f, for the transmission path ij, involves the flanking element i with a surface area S_i in the sending (i.e. source) room and the corresponding flanking element j in the receiving (i.e. receiver) room with a surface area S_j, and hence is defined as,

$$(R_f)_{ij} = R_{ij} = 10 \lg \left[\frac{1}{\tau_{ij}}\right] = 10 \lg \left(\frac{W_D}{W_{ij}}\right) = 10 \lg \left(\frac{I_i S_D}{I_j S_j}\right) \tag{3.59}$$

The quantity W_D is defined as the sound power incident on the main partition and W_{ij} is defined as the radiated power from element j in the receiving room which is caused by vibration transmission from the element i from the sending room. The sound intensity I_i at the wall elements is assumed to be the same at all surfaces in the sending room as an outcome of the diffuse sound field approximation, however, the intensity I_j is radiated from the element j in the receiving room from j^{th} wall element and might be varied according to the wall element radiation factor. The diffuse field approximation is a good approximation for simplified calculations of the direct as well as flanking sound transmissions between adjacent rooms, however, there are many discrepancies which remained in the prediction models that we discuss in sound filter design chapter (Chapter **4**).

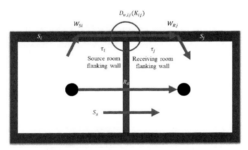

Figure 3.4: *Sound transmission paths: Direct and flanking paths*

The sound powers W_D and W_{ij} defined above now can be expressed in terms of mean squared sound pressure over the respective surface areas of the wall elements as given in Equation **3.60**.

$$W_D = \frac{\langle \bar{p}_S^2 \rangle}{4\rho_0 c_0} S_D \text{ and } W_{ij} = \frac{\langle \bar{p}_R^2 \rangle_{ij}}{4\rho_0 c_0} A_R \tag{3.60}$$

The flanking sound reduction index for the path ij can be written in the form of the following expression given in Equation **3.61**.

$$R_{ij} = L_s - (L_R)_{ij} + 10 \log \frac{S_D}{A_R} \qquad (3.61)$$

To include the properties of the flanking elements into R_{ij}, the radiated sound power W_{ij} of the pertinent element j on the receiving room side must be represented in terms of radiation factor (i.e. radiation efficiency σ). It means that W_{ij} can be expressed by $W_{ij} = \rho_0 c_0 S_j \langle \tilde{u}_j^2 \rangle \sigma_j$, which gives the sound pressure as,

$$\langle \tilde{p}_R^2 \rangle_{ij} = \frac{4\rho_0^2 c_0^2 S_j \langle \tilde{u}_j^2 \rangle \sigma_j}{A_R} \qquad (3.62)$$

A corresponding equation may be found for the sending room, linking the sound pressure level and the velocity u_i of the flanking element thereby using the transmission factor τ_i of the flanking element. Hence,

$$\tau_i = \frac{W_t}{W_i} = \frac{\rho_0 c_0 S_i \langle \tilde{u}_i^2 \rangle \sigma_j}{W_i} = \frac{\rho_0 c_0 S_D \langle \tilde{u}_i^2 \rangle \sigma_j}{W_D} \qquad (3.63)$$

Where W_t and W_i denote the transmitted and incident power on the flanking elements i and j respectively. In the last expression of Equation **3.56** we have made use of the fact that the sound intensity is the same everywhere at all surfaces in the sending room. Using the expression for W_D from Equation **3.60** we get,

$$\langle \tilde{p}_S^2 \rangle = \frac{4\rho_0^2 c_0^2 \langle \tilde{u}_i^2 \rangle \sigma_j}{\tau_{ij}} \qquad (3.64)$$

From Equation **3.57**, Equation **3.60**, Equation **3.62** and Equation **3.64**, we get the transmission loss for the ij flanking path given below.

$$\tau_{ij} = \tau_i . \frac{\langle \tilde{u}_j^2 \rangle}{\langle \tilde{u}_i^2 \rangle} \left(\frac{\sigma_j}{\sigma_i} \right) \left(\frac{S_j}{S_S} \right) \qquad (3.65)$$

and the corresponding sound reduction index for this path as,

$$R_{ij} = R_i + 10 \lg \left(\frac{\langle \tilde{u}_i^2 \rangle}{\langle \tilde{u}_j^2 \rangle} \right) + 10 \lg \left(\frac{\sigma_i}{\sigma_j} \right) + 10 \lg \left(\frac{S_S}{S_j} \right) \qquad (3.66)$$

The second term in right side of Equation **3.66** gives us the velocity difference denote by $D_{v,ij}$ and by following the ISO 12354-1 [6] we can define the flanking sound reduction index as a mean value from measurements in two directions exchanging the sending and receiving rooms. Therefore, we can write,

$$\bar{R}_{ij} = \frac{R_i + R_j}{2} + \bar{D}_{v,ij} + 10\lg\left(\frac{S_S}{\sqrt{S_i S_j}}\right) \tag{3.67}$$

with the direction-averaged velocity level difference given by $\bar{D}_{v,ij} = \frac{D_{v,ij} + D_{v,ji}}{2}$.

However, the applications of Equation **3.67** are rather limited since it is only valid for heavy constructions where the forced transmission can be neglected. This can have an implication (or limitation) in auralization model development, however, the main focus of auralization model is the sound transmission for adjacent rooms and since in adjacent rooms case the transmission from main partition is more dominant than the flanking transmission, therefore, this discrepancy may be ignored.

3.4. Combining Direct and Flanking Transmissions

As discussed in the previous sections that the determination of the flanking transmission is a complicated process. A series of international standards has been developed for laboratory measurements for the prediction of flanking transmission. The velocity level difference across a junction, which is introduced in above section is not an invariant quantity as it depends on the actual energy losses in the receiving element. In this section, we discuss the bending waves across the junctions and the type of the junctions which is an important part in determination of overall sound transmission from source to receiver at some point in the receiving room. While doing so it is then quite helpful to find predict the level difference between the adjacent rooms and further proceeding toward auralization of the sound insulation.

3.4.1.   Bending wave transmission across plate intersections

A plane bending wave is incident on an intersection involving three plates or four plates. All plates are assumed to be of infinite extent and a bending wave in the plate of thickness h_1 is assumed to be incident normally to the axis of the intersection. Two auxiliary quantities are introduced as $\psi = \sqrt{\frac{m_2 B_2}{m_1 B_1}}$ and $\chi = \frac{k_{B2}}{k_{B1}} = \sqrt[4]{\frac{m_2 B_1}{m_1 B_2}}$, which are the ratios of the impedances and wavenumbers of the actual plates, respectively. Where, m_1 and m_2 are the mass per unit surface area and B_1 and B_2 are the bending waves stiffness of the plates. In practice, it is normally assumed that the plates are in a line and have identical material and thickness. With this assumption one may express the sound reduction indexes using a single parameter, which is the ratio of these auxiliary quantities, such as,

$$\frac{\psi}{\chi} = \frac{c_{B1} B_2}{c_{B2} B_1} \tag{3.68}$$

Where c_{B1} and c_{B2} are the phase speeds of the waves travelling in plates. If the plates have identical material properties, we can further find that this ratio is given by the thickness ratio of the plates given by $\left(\frac{\psi}{\chi}\right) = \left(\frac{h_2}{h_1}\right)^{\frac{5}{2}}$. The sound reduction indexes R_{12} and R_{13} for a T-junction are then given by,

$$R_{12} = 20\lg\left[\sqrt{\frac{2\chi}{\psi}} + \sqrt{\frac{\psi}{2\chi}}\right] \quad \text{and} \quad R_{13} = 10\lg\left[2 + 2\left(\frac{\psi}{\chi}\right) + \frac{1}{2}\cdot\left(\frac{\psi}{\chi}\right)^2\right] \tag{3.69}$$

The corresponding expressions applied to a T-junction are then given by,

$$R_{12} = 20\lg\left[\sqrt{\frac{\chi}{\psi}} + \sqrt{\frac{\psi}{\chi}}\right] + 3\,\text{dB} \quad \text{and} \quad R_{13} = 20\lg\left[1 + \left(\frac{\psi}{\chi}\right)\right] + 3\,\text{dB} \tag{3.70}$$

3.4.2.   Vibration reduction index K_{ij}

An invariant quantity, which characterize the transmission across a junction of finite element under diffuse sound field assumptions defined in ISO 12354-1 [6], is being called vibration reduction index having the symbol K_{ij}. For a complete picture we start with classical calculations concerning bending wave transmission across plate intersections involving three junctions (i.e. T-junction, X-junction and L-junction). For detailed derivation refer to [40]. This vibration reduction index K_{ij} characterizes the transmission across a joint between finite size elements under diffuse field conditions. This index is determined by measurements by taking space time averaged velocities and structural reverberation time of the actual elements. The structural reverberation time determines the damping of the elements and is expressed by the equivalent absorption length a_i of the element i and hence the vibration reduction index in given as,

$$K_{ij} = \overline{D}_{v,ij} + 10 \lg \left[\frac{l_{ij}}{\sqrt{a_i a_j}} \right] \tag{3.71}$$

Here, l_{ij}, is the length of the junction between elements i and j. The relationship between reverberation time T and absorption length a is given by Equation **3.72**. The reference frequency f_{ref} is chosen to be **1000 Hz**.

$$a_i = \frac{2.2\pi^2 S_i}{c_0 T_i} \sqrt{\frac{f_{ref}}{f}} \tag{3.72}$$

Here, one important point to be mentioned that a relationship between K_{ij} and a transmission factor (or reduction index) can be established based on bending wave power in conformity with Cremer's definition [**40**] or more recently based on relationship between the vibration reduction index K_{ij} and the coupling loss factor η_{ij} according to [**41**]. Basic junctions are show in Figure **3.5**.

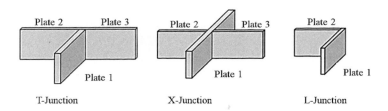

Figure 3.5: *Types of junctions* [**15**]

The calculations of absorption coefficient α_k, for a structural element i (e.g. Plate 2 for T-Junction) can be deduced from the vibration reduction index K_{ij}, according to ISO 12354 [**6**], at the junction between the considered elements i and j connected to it by the relation given in Equation **3.73**. Here, j are all those elements (e.g. Plate 1 and Plate 3 for T-Junction in Figure **3.5**) which are connected to the considered element i at its boarder k.

$$a_k = \sum_{j=1}^{3} \sqrt{\frac{f_{c,j}}{f_{ref}}} 10^{-\frac{K_{ij}}{10}} \qquad (3.73)$$

In Equation **3.73**, the method adopted to calculate K_{ij} is based on ISO: 12354-1 (Annex E) [**6**]. In this annex, buildings with heavy structures such as masonry or concrete wall and floors are considered, for which junctions are characterized by K_{ij}. The presence of lightweight elements such as partitions, separating walls or façades is still possible, thus leading to junctions mixing heavy and lightweight elements and also treated using K_{ij}. In [**6**] two types of K_{ij} data are introduced: **1**) the empirical data deduced from standardized measurements or theory and **2**) data from simulations, which is more difficult to use with additional input parameters, however, more traceable. Several types of structural joints are considered in [6] to calculate the empirical data on K_{ij}. For the detail we refer to ISO: 12354-1 (Annex E). The ISO [6] considers the flanking transmission between two elements and the junction between them only. However, while the individual higher order flanking paths may be insignificant compared to the direct transmission, the sum of the contributions of the higher order paths may be significant [**79**]. The exclusion of higher order paths limits the ISO:12354-1 [**6**]. Which means that this method is only capable to calculate the apparent sound reduction index between adjacent rooms. The rooms that do not share the common elements of junction require the use of higher order flanking paths and complex SEA model would be required. As the scope of dissertation is restricted to adjacent rooms for sound insulation filters construction (for indoor case), that is why ISO:12345-1 [**6**] (Annex E) is used for determination of the input data for the flanking transmission across the adjacent rooms.

3.4.3. Combining Multiple Surfaces

When calculating the sound reduction index of a partition consisting of an assembly of two or more parts or surfaces, one normally assumes no interaction between the different parts; each part vibrates independently driven by the incident sound pressure. This is certainly a simplification but it may be justified by giving a rough and reasonable estimate. The total transmission factor τ_{tot} of n number of partial surfaces S_n having transmission factor τ_n will be given by

$$\tau_{tot} = \frac{\tau_1 S_1 + \tau_2 S_2 + \cdots + \tau_n S_n}{S_1 + S_2 + \cdots + S_n} = \frac{1}{S_o} \sum_{i=1}^{n} \tau_i S_i \qquad (3.74)$$

S_o is the total area. Expressed by the corresponding sound reduction indices, we get

$$R_{tot} = 10 \lg \left[\frac{S_o}{\sum_{i=1}^{n} S_i 10^{-\frac{R_i}{10}}} \right] \qquad (3.75)$$

An example showing the use of this expression is given in Figure **3.6** (Figure 9.1 [**20**]), where there are just two components ($n = 2$) giving a diagram useful for dimensioning a partition containing a door or window. It should be noted that R_0 is the sound reduction index belonging to the total area S_o, i.e. the index before the smaller part of area S_1 with reduction index R_1 is inserted. The explicit expression, certainly assuming, $S_1 \leq S_0$, is

$$R_0 - R_{total} = 10 \lg \left[1 - \frac{S_1}{S_0} + \frac{S_1}{S_0} \cdot 10^{\frac{(R_0 - R_1)}{10}} \right] \qquad (3.76)$$

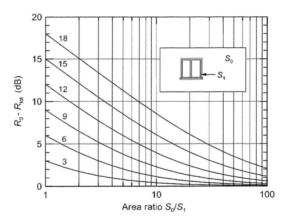

Figure 3.6: *Calculation of the sound reduction index of a composite construction, e.g. a partition containing windows [**20**]. $R_o - R_1$ is the difference in sound reduction indices*

The equations summarized in Chapter **3** are the basis for sound insulation prediction according to standardized procedures and latest available research. In residential and worksite premises, especially in the urban areas, the international standards provided by ISO have reflected the trends in growing annoyance due to the indoor/outdoor noise and indoor background speech. These trends are increasing both in number and covering broader aspects. However, the guidelines do not provide the optimal acoustic satisfaction and intern the accurate evaluation of building performances when specific sounds e.g. conversation varying in intelligibility originate from the adjacent office or a transient noise from outdoor moving sound source, cause the disturbances in daily life's physical and mental work. Therefore, measurement procedures applied in laboratory or in the field are just one part of the story. There are several standards and available methods, as discussed in Chapter **3**, which describe the performance of building elements in terms of noise reduction and level reduction indices in the form of a single number value and/or frequency dependent curves. Nevertheless, it can be assumed that these quantities are insufficient to describe the real situation for the perceptual evaluation of noise and comfort.

Hence, it is desirable to develop advanced insulation prediction techniques and auralization tools that simulates the sound field at listener's ears from predicted or measured data using auralization of these noises with better subjective impressions, and psychoacoustic and psychological factors. The basic principle of building acoustic auralization is to simulate the alteration of a sound signal from its source to the receiver end via transmission through the elements. The auralization of a situation where either the speech spoken in one office or the noise produced by an outdoor source is transmitted through building structures, requires the sound source modelling, sound propagation (e.g. from adjacent room or from outdoor etc.) and its transmission through wall elements, and the insulation characteristics of the direct and flanking parts of the dwellings.

In order to auralize these situations for office-to-office situations or for insulation against the outdoor sounds, both the level and the spectral characteristics of sources are highly dependent on the sound insulation curves of the building constructions which are separating the source and receiver. The fundamentals of sound insulation prediction according to ISO and latest research are summarized in Figure **3.7**. It contains the input data characterizing the features of the building elements, and it shows the combination of the performance of the building elements into the final apparent sound reduction of building environments constructed from the elements.

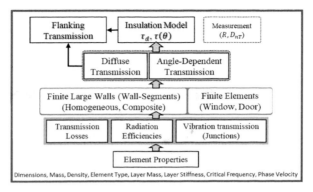

Figure 3.7: *Flow diagram to compute angle dependent and diffuse sound transmission [80]*

In the next chapter, the standard sound insulation quantities are used as input data for signal-theoretical techniques finally leading to filter implementations for real-time audio signal processing.

4

Sound Insulation Filters: Auralization

This chapter explains the design procedures for airborne sound insulation filters based on the knowledge of performance of building elements and characteristics of sound transmission through these elements (i.e. from sound insulation metrics either predicted or measured in a laboratory). The sound insulation filters are developed for office-to-office sound transmission (i.e. adjacent rooms) and sound transmission from outdoor sound sources against outdoor environment (e.g. façade sound insulation). The prerequisite for sound insulation filter design is accurate sound insulation input data which is supposed to be based on the sound insulation prediction theory as well as on available laboratory measurements according to international standards. Most of the input data for auralization is available in compliance with standardized data formats of sound insulation prediction (such as ISO 12354-1 [6] and ISO 12354-3 [17]) from where point to point transfer functions from source to the listener are possible to calculate. Nevertheless, there are several simplifications in ISO prediction models and available approaches which are discussed in Chapter 1 along with their limitations.

It is inevitable to address these limitations in prediction models during transfer function calculation and to achieve plausible auralization and plausible reproduction of level and coloration. The compensation of these limitations is achieved by introducing advanced models which are intended to include room acoustical simulations of the sound eld in the source and receiving rooms, and modelling of the excitations and radiations of incident waves on the wall elements. Subsequently, the input data from these models can be used to simulate airborne sound transmission paths from source to receiver placed at some random location in the building. Furthermore, certain improvements in these models are also introduced which deal with reverberant sound fields in both source and the receiving rooms including room impulse response synthesis based on one-third octave band values of the reverberation time. We introduce new approaches which bring certain advancements in sound insulation predictions and filters design, for example, consideration of non-diffuse sound field, angle dependent transmission, distributed vibration energy patterns on the building wall elements of source and receiving rooms and transmission of energies through direct as well as flanking elements. Diffuse field theory work very well for finite monolithic walls, however, if the walls sizes are large enough the diffuse field assumptions fail because the sound transmission critically dependents on incident angle of the plane wave on the large finite walls.

To achieve these advancements in sound insulation predictions and filters construction, in first place, we introduce sound source directivities, the source and receiving room reverberation, which depend on the room characteristics (e.g. room volume, absorption), and the spatial variation of the sound field inside source room and on its boundaries (i.e. walls). Secondly, the sound insulation transfer functions from source to the receiving room are elaborated for extended (and large) walls by using the concept of dividing the individual building element into a multitude of secondary sound sources (i.e. finite-sized patches) and considering the angle-dependent transmission. Commonly, wall elements of the dwellings are either homogeneous elements (e.g. a single homogeneous wall element) or consist of an assembly of two or more parts or surfaces (i.e. doors, windows). This chapter starts with the filters construction for the adjacent rooms and outdoor sound sources, and explains the corresponding real-time algorithm techniques in subsequent sections. The fundamentals of auralization techniques are discussed that are used for the final binaural reproduction at the listener's end in the receiver room.

The auralization technique consists of a signal-filter model and its implementation in the audio signal processing domain [9]. The time signal at a receiver in a room can be calculated from the source signal and the transfer function from source to the listener. The source plays a "signal", which is convolved with an impulse response filter in order to change the signal by imprinting the sound path properties and by reproducing the resulting signal to a human listener. In the next sections, it is described how to obtain filter impulse responses.

4.1. Filters for Adjacent Rooms: Simplified Approach

The most commonly used sound insulation prediction methods are based on the equations presented in Chapter **3** which refer to ISO [**6,17**] and available research [e.g. **48,49,50,51**]. Specifically, SEA (Statistical Energy Analysis), ASEA (Advanced Statistical Energy Analysis) and FEM (Finite Element Methods) are used when higher precision in the results is required. It is very challenging to assume that which one of the available methods is accurate and precise for the purpose of sound insulation filters and auralization. In reality, each method has certain limitations depending on the assumptions made during their development and on assumption on the source signal which can be stationary, transient, dominated by low or high frequencies, etc. In fact, also measured transmission coefficients for building elements of direct and flanking transmission can be used as basis for filter design. As the origin of sound insulation data is flexible concerning the filter design, in the following, the approach of ISO [**6,17**] is applied. It is widely used in most of the European countries which are based mainly on publications of Gerretsen [**18,42**] and on refinements from Vigran [**20**], Rindel [**22**], Fahy [**23**] and Davy [**25**]. As an established method, the ISO [**6**] is chosen as the foundation in simplified approach of the sound insulation prediction model for filter designs and to build an auralization for adjacent rooms. The first application of this kind of airborne sound insulation auralization was introduced by Vorländer and Thaden [**3,10,14**], who implemented Gerretsen's prediction method [**18,42**] in the signals and filters domain and presented an auralization of airborne sound insulation using binaural technology through headphones. The basics of their approach are as follows.

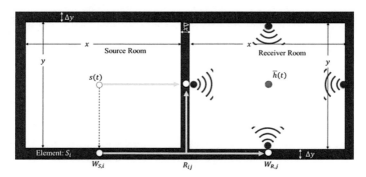

Figure 4.1: *Typical adjacent source and receiving rooms with receiving room walls represented as secondary point sources radiating elements*

As discussed in Chapter **3**, the standardized sound level difference is expressed by the transmission coefficient $\tau_{ij} = 10^{-0.1R_{ij}}$ of the transmission path ij between two rooms which is given in Equation **4.1**. Here i and j denote the source and receiving room wall elements, respectively, for the transmission path ij, with receiving room volume V and the separating (direct) element area S_D between the two rooms. In Figure **4.1**, a typical adjacent source and receiver rooms, separated by wall element are shown.

$$D_{nT} = -10 \log \sum_{\forall ij} \tau_{ij} + 10 \log \frac{0.32V}{S_D} \tag{4.1}$$

The resulting average sound pressure level in the receiver room is calculated for all transmission paths by Equation **4.2**. By introducing the sound energy signals p_S^2 and p_R^2 as mean squared energies in the source and receiving room respectively, Equation **4.2** is expressed in energetic form and is given in Equation **4.3**.

$$L_R = L_S + 10 \log \sum_{\forall ij} \tau_{ij} + 10 \log \left(\frac{S_D}{0.32V} \frac{T}{0.5} \right) \tag{4.2}$$

$$p_R^2 = p_S^2 \frac{S_D}{0.32V} \frac{T}{0.5} \sum_{\forall ij} \tau_{ij} \tag{4.3}$$

In these equations, the sound energy radiating elements i.e. the receiving room walls (denoted as secondary sources (SS) in [**3,10,46,47,80**]) are approximated as point sources located at the centres of the wall elements and are representing the whole vibration wave pattern on the wall as a single point source. The balance between direct and reverberant part of sound field is very important in perception of the spatial characteristics of the rooms. This energy balance is obtained by using definition of reverberation distance r_{rev}, defined in Equation **2.4** of Chapter **2**, and is computed through the ratio of energies given by relationship, $\frac{E_{rev}}{E_{dir}} = \frac{16\pi r^2}{A}$. The quantities, E_{dir} and E_{rev} are the energies of direct and reverberant sound field at a distance r from the secondary source. The A is the equivalent absorption area of the receiving room. For uncorrelated direct and reverberant sound fields, the contribution of the transmission path ij to the mean squared pressure in terms of the reverberant and the direct field can be written as $p_{R,ij}^2 = p_{R,ij,dir}^2 + p_{R,ij,rev}^2$ by using Equation **2.5** of Chapter **2**. The temporal effects of the receiving rooms in terms of reverberation are included by simulating its impulse response, $h(t)$. Here, at first, the direct sound is removed from the impulse response as it is already included in the transmission path calculation in its binaural form $HRIR\left(t - \frac{r_j}{c_0}, \theta_j, \varphi_j\right)$. Subsequently, it is approximately equalised to a white spectrum and normalized in energy [47,80]. The time domain representation of the binaural signal from source to receiver of the transmission path ij is given in Equation 4.4. All binaural contributions from the radiating wall elements (i.e. secondary sources) are summed up to get the final signal. The impulse responses in this context are shown in Figure 4.2 (left) and Figure 4.2 (right) in frequency and time domain, respectively. The algorithmic process chain for auralization of adjacent rooms is given in Annex A.

$$p_{R,ij}(t) = \sqrt{\frac{S_D}{0.32V}\frac{T}{0.5}\frac{\tau_{ij}}{16\pi r_{ij}^2 + A}}\ p_s(t) * \left[\sqrt{A}\ HRIR\left(t - \frac{r_{ij}}{c_0}, \theta_j, \varphi_j\right) + \sqrt{16\pi r_{ij}^2}\ h(t)\right] \quad (4.4)$$

This model forms the foundation of sound insulation auralization of typical rectangular adjacent rooms as proposed in [3,10]. However, this model incorporates several simplifications as can easily be observed in Equation 4.4. These simplification are: at first, the transfer functions (τ_{ij}) between i^{th} element of the source room and j^{th} element of the receiver room are valid for point to point transmission only, although extended wall elements are present in real built structures. These extended walls may be homogeneous single elements or consist of an assembly of different building elements such as; door and portals. Secondly, in receiving room it is assumed that the sounds are apparently radiated from one point representing the whole vibration wave pattern of the wall elements. Although, the spectrum of these point radiations are exact, yet the wave pattern on the walls are replaced by a point source at the centre of the wall. Thirdly, the transfer functions (i.e. room impulse responses) which include the room acoustics of the receiving room are assigned to the same point of radiations instead of using separate room impulse responses for each source receiver combination (i.e. radiating point sources on walls and the listener).

Another important aspect is source directivity that is neglected, which might reasonably contribute toward a specific distribution of sound intensity on the surfaces of the source room walls. The amount of transmitted energy would be different for different paths, in particular if sources are placed close to walls (such as loudspeakers or TV sets). In this chapter the basic sound insulation auralization model is the starting point. All simplifications in this model are intended to be address and discussed in next sections of this chapter. As a result, the advanced sound insulation model and filters for auralization are presented.

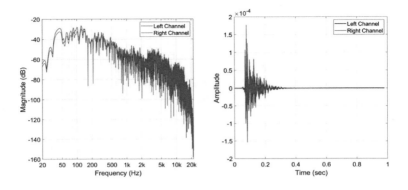

Figure 4.2: *Final binaural impulse responses at the listener end: Frequency domain (**left**) and time domain (**right**)*

4.2. Filters for Adjacent Rooms: Extended Approach

In Section **4.1**, it was discussed auralization of sound insulation filters for simple adjacent rooms which is the starting point of the advanced auralization models [**80**]. In this extended approach, in first place, it is taken into account the source room acoustics by considering a more complex sound field incident on the source room walls consisting of a direct and a diffuse field component. This was already introduced by Rodríguez-Molares [**64**], who included the temporal decay of the room responses. The sound energy transmitted via direct and flanking paths to the adjacent receiving room is now specific for all surface elements, depending on the sound pressure hitting the corresponding wall elements in the source room due to source position and directivity.

Secondly, the influence of reverberation of source and receiving rooms and the balance between direct and reverberant energies inside the receiving room are incorporated into sound insulation transfer functions. These transfer functions are developed for extended radiating walls (i.e. the receiving room wall elements) by using a grid of point sources (known as secondary sources). Thirdly, the angle dependent sound transmission is considered for large composite and homogeneous walls. It is also adopted a procedure to synthesize the room impulse response $h(t)$ from the reverberation time T to include the effects of absorption of room boundaries as well as to simulate plausible real rooms. In the next sections we will discuss each aspect of extended approach for sound insulation prediction model.

4.2.1.   Sound Source Directivity

Generally, the diffuse sound field conditions are assumed for the prediction of sound insulation metrics in adjacent rooms. However, the sound source directivity in the source room might have a significant influence on calculate the sound field in the source room and, in turn, on the energy transmitted to the receiver room. The energetic source directivity Q_s is introduced for computing the sound energy distribution in the source room and at its direct and flanking wall elements. As an example, it is illustrated the directivities of a trumpet and a typical loudspeaker as shown in Figure **4.3**. The directivities are normalised to guarantee that $\int_s Q_s(\theta, \varphi) = 4\pi$ for $\theta = [0, \pi]$ and $\varphi = [0, 2\pi]$, which is the approach adopted by [**64**].

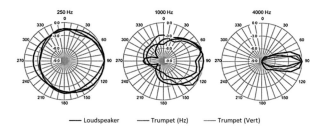

Figure 4.3: *Loudspeaker (typical HiFi system in common rooms) directivity and its comparison with trumpet directivities; derived from* [**43**]

4.2.2.   Room Impulse Response Synthesis

The computation of room impulse response denoted in this thesis as $h(t)$, is based on the reverberation time T of the source room and artificial noise representing the sum of the room reflections [**9,21**]. Under diffuse-field conditions, the sound energy at the surfaces of the source room wall elements is considered equal for each wall. In the extended approach, the synthesis of source room impulse responses to the surface elements of the walls is necessary to include the effects of absorption of room boundaries as well as to simulate the cases where an equivalent real room is not present. It means that we require room impulse response $h(t)$ at any point in the source room (especially near the surface of the each wall element) from where we can estimate sound energy at a particular wall surface [**80**]. The approximated $h(t)$ is obtained through a linear combination of filtered exponential decay signals. Consider a time domain signal $g(t, T)$, as given in Equation **4.5**, with $n(t)$ as a normally distributed time domain random variable having zero mean and unit standard deviation. The signal $g(t, T)$ decays 60 dB for each T, for all frequency bands [64].

$$g(t, T) = \sqrt{\frac{13.81}{T}} e^{-\frac{6.91t}{T}} \cdot n(t) \qquad (4.5)$$

The factor $\sqrt{\frac{13.81}{T}}$ in Equation **4.5** normalizes $g(t, T)$ in energy. From the linear combinations of filtered signals $g(t, T)$, the impulse response $h(t)$ is synthesized, which then decays at different rates for each frequency band (given in Equation **4.6**).

$$h(t) = \sum_{\forall k} \alpha_k \cdot g(t, T_k) * F_k(t, k) \qquad (4.6)$$

Here, T_k is reverberation time and the function $F_k(t, k)$ is a set of band-pass filters in time domain for each kth one-third octave band. As mentioned in [64] The function $g(t, T)$ tends to a white spectrum because of a convolution of the Fourier transform of $e^{-\frac{6.91t}{T}}$ and a white spectrum of $n(t)$. However, there appear slight variations in the spectrum of $h(t)$ due to the statistical nature of $n(t)$, that must be compensated by α_k given in Equation **4.7**.

$$\alpha_k = \frac{\sqrt{0.23 f_k}}{\sqrt{\int_{-\infty}^{\infty} \left(g(t, T_k) * F_k(t, k) \right)^2 dt}} \qquad (4.7)$$

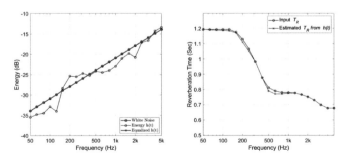

Figure 4.4: *(a) Example of a spectrum of synthesised $h(t)$, (b) Computed reverberation time in comparison with input data*

In Figure **4.4**, the reverberation time for the impulse response $h(t)$, is obtained using ISO 3382 with the one-third octave band values, T_k. Both figures show the reproduction of a room impulse response with the required properties and the results are quite similar to that reproduced by Rodriguez-Molares [64]. This approach for room impulse response synthesis is extended by one step more for our purpose.

4.2.3. Sound Field in the Source Room

In closed spaces, it is assumed that the direct sound field propagates and decay with time as in free-field conditions. The reverberant sound field is evenly distributed throughout the space. This phenomenon is described in classical sound field theory for sound propagation in rooms given by Equation **4.8** in the simple form (derived from Chapter **2**). Let us consider two simple rectangular adjacent rooms, with given dimension and separated by main partition, as shown in the Figure **4.5**.

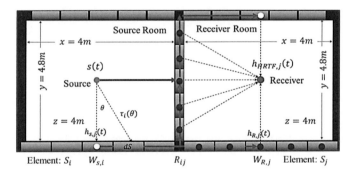

Figure 4.5: *Adjacent source and receiver rooms with wall elements as multitude of secondary source [80]*

A sound source with specific directivity pattern is placed at an arbitrary position the source room. The directivity of this source is represented by Q_s and its sound power level by L_w. The source produces a sound pressure level L_s at distance r inside a source room with an equivalent absorption area A_s, and is given by Equation **4.8**.

$$L_s = L_w + 10\log\left(\frac{Q_s}{4\pi r^2} + \frac{4}{A_s}\right) \tag{4.8}$$

Equation **4.8**, inherently incorporates the influence of the source room reverberation, the directivity of the source, and the same balance between direct and reverberant energy as considered in previous approaches [10,14,47].

Let the acoustics power of the source is $W_a = 10^{-12}10^{-0.1L_w}$ (source acoustic power in Watts). The mean squared sound pressure at any point inside the source room in energetic notation can be calculated by Equation **4.9**,

$$p_s^2 = \rho_o c \, W_a \left(\frac{Q_s}{4\pi r^2} + \frac{4}{A_s} \right) \tag{4.9}$$

To calculate the sound pressure, in signals and filter domain, at any point in the source room a loudspeaker with directivity Q_s as shown in Figure **4.3** is selected as an example sound source to analyse the influence of the source directivity on the transmitted energy to the receiving room walls for the direct as well as for flanking paths. Note that, as shown in the Figure **4.5**, the extended model includes wall elements which are further subdivided into small segments known as "patches", instead of computing the power incidence on the wall as a whole. Let $s(t)$ be source signal normalized in power and $h_{s,i}(t)$ is energetically normalised impulse response of the source room calculated at a distance r_i from it by using Equation **4.9**. Let the directivity of source in the direction of receiver point on wall is Q_s, the time domain representation of Equation **4.9** is rewritten in the form of Equation **4.10** [47,80].

$$h'_{sp,i}(t) = \sqrt{\rho_o c W_a} \left(\sqrt{\frac{Q_s}{4\pi r_i^2}} \, \delta \left(t - \frac{r_i}{c} \right) + \sqrt{\frac{4}{A_s}} \, h_{sp,i}(t) \right) \tag{4.10}$$

From Equation **4.10**, it is possible to calculate the sound pressure inside the source room on any point on the surface of wall elements. This means that the incident sound pressure and hence sound energy on each segment (i.e. "patch") can be calculated. If the building wall elements consist of an assembly of components such as doors and portals, the sound pressure at each component is feasible to calculate for which the algorithm flow chart is given in Annex **A.3**.

After calculating $h(t)$ from the energetically normalized impulse response, at first the direct sound and the first part of the exponential decay are removed from this impulse response as the direct sound is already included in the transmission path calculation. The gap between the direct sound and the reverberation part ($h(t)$) can be as wide as given by the mean free path ($\overline{d} = \frac{4V}{S}$), which is the averaged distance travelled by a sound between reflection. Here, V is the volume and S is the surface area of the room [44]. Subsequently, it is equalized to white spectrum and normalised in energy. The effect of α_k in the frequency domain can be seen in Figure 4.4. The resulting impulse response is denoted $h(t)$. It contains the room response without the direct sound which arrives at 4.4 ms, whereas, the first reflection arrives at 7.5 ms. The algorithm flow implemented for synthesised $h(t)$ is shown in Annex A.2.

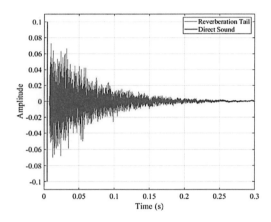

Figure 4.6: *Example of a synthetic room impulse response (reverberation tail after first part extracted)*

4.2.4.   Incident Sound Energy at Wall Surface (Source Room)

In reverberant rooms, as a general case, the diffuse sound field is a good approximation while dealing with the sound field propagation, and the results for stationary conditions and sound decays might be applied to measure the sound power of a source. A rather simple modification to the stationary sound field is to separate the direct sound from the reverberant part of the impulse responses (i.e. reverberation tail) as proposed by [3,10,14]. Hence, the incident sound power on any wall element i in the source room with surface area S_i is taken as a combination of direct and the diffuse sound fields. In real dwellings the walls (specially the outer wall elements) are usually consists of an assembly of two or more components or surfaces; such as doors, portals and windows, which are known as composite walls. For this reason, we divide the wall elements into components (patches). Let's consider a patch on the wall element i, with surface area $S_{p,i}$. Under diffuse sound field conditions the reverberant part of the incident sound power $W_{s,rev}$ on any patch of any wall element is given by Equation **4.11**.

$$W_{sp,rev} = \frac{p_{s,rev}^2 \cdot S_{p,i}}{4\rho_0 c_0} = \frac{W_a \cdot S_{p,i}}{A_S} \qquad (4.11)$$

On the other hand, under free-field conditions, the incident direct sound power $W_{sp,dir}$ is calculated on this patch and is given in Equation **4.12**.

$$W_{sp,dir} = \frac{W_a}{4\pi} \int_{S_{p,i}} \frac{Q_{s,p}}{4\pi r_{p,i}^2} |\cos\theta_{p,i}| dS_p \qquad (4.12)$$

In Equation **4.12**, $Q_{s,p}$ is representing the source directivity in the direction of this patch, and $r_{p,i}$ and $\theta_{p,i}$ are the distance and the incidence angle from the source point to the infinitesimal patch area dS_p. These quantities ($Q_{s,p}$, $r_{p,i}$ and $\theta_{p,i}$) depend on the room geometry. If the integral in Equation **4.12** is represented by Equation **4.13**, then by combining Equation **4.12** and Equation **4.13** the incident sound power on a single patch p of wall element i results in the form of Equation **4.14**.

$$F_{p,i} = \int_{S_p} \frac{Q_{s,p}}{4\pi r_{p,i}^2} |\cos\theta_{p,i}| dS_p \qquad (4.13)$$

$$W_{sp,i} = W_a \left(\frac{F_{p,i}}{4\pi} + \frac{S_{p,i}}{A_S} \right) \qquad (4.14)$$

The integral $F_{p,i}$ is approximated numerically for not very large patches and in not very close positions to the walls as introduced by [64] and it is even more appropriate after the wall has been subdivided into small segments (patches), thus relaxing the condition of constant conditions on the surface. This integral is obtained by assuming that $Q_{s,p}$, $r_{p,i}$ and $\theta_{p,i}$ do not vary significantly along, $S_{p,i}$, therefore, these factors are taken out of the integral and approximate solution of Equation **4.13** is now given in Equation **4.15**.

$$F_{p,i} \approx \frac{S_{p,i} Q_{s,p}}{4\pi r_{p,i}^2} |\cos\theta_{p,i}| \qquad (4.15)$$

The vector $r_{p,i}$ is the distance from the source to the centre of patch p with an incidence angle $\theta_{p,i}$ and $Q_{s,p}$ denotes mean directivity value in the direction $\theta_{p,i}$. In this method, the integral $F_{p,i}$ is calculated by the adaptive Simpson's integration method. The incident power on patch p given in Equation **4.14**, is now represented by its corresponding instantaneous incident sound power in time domain as follows.

$$W_{sp,i}(t) = W_a \cdot s(t) * \left(\sqrt{\frac{F_{p,i}}{4\pi}} \cdot \delta\left(t - \frac{r_{p,i}}{c}\right) + \sqrt{\frac{S_{p,i}}{A_s}} \cdot h'_{sp,i}(t) \right) \tag{4.16}$$

In Equation **4.16**, $s(t)$ is the source signal normalized in power and $h_{sp,i}(t)$ is energetically normalised impulse response of the source room for patch p of wall element i, from which the direct sound is removed. The synthesis of source room impulse responses at the surfaces of the patches is necessary to include the temporal effects of the source room, the effects of absorption of room boundaries as well as to simulate the cases where an equivalent real room is not present. The method for calculation of room impulse response is described in Section **4.2.1** and the algorithm flow chart is given in Annex **A.3**.

4.2.5. Sound Transmission

The incident sound power on each wall element (or patch) of the source room is transmitted to receiving room via main partition (i.e. separating wall element) as well as through flanking paths connected by junctions. In this section, we discuss both direct and flanking sound transmission separately. The sound transmission depends on two main factors which are **1)** the dynamic response and **2)** the radiation factor (efficiency) of the building elements. The sound radiations from the plates are due to both forced and resonant vibrations as discussed in Chapter **2**. Direct sound transmission (i.e. Direct-to-Direct (**Dd**) transmission path) through separating wall element between the adjacent rooms is relatively easy to compute as compared to sound transmission for flanking paths which is relatively a more complicated case.

4.2.5.1. Direct Sound Transmission

As discussed in Chapter **3**, the sound transmission coefficient of the direct transmission (from source to receiver via main partition) between adjacent rooms is a ratio between the incident and transmitted powers, therefore, it is necessary to compute the transmission coefficients for the partition between these adjacent rooms at first. Generally, the transmission coefficients are estimated based on diffuse field assumptions for which the spatially averaged values of transmission coef cients are commonly used for prediction of sound insulation and on taking transmission coefficients from D_{nT} or R_{ij} data as given in Equation **4.1**. Other important sound insulation metrics such as; vibration velocities on the surface of elements, radiation efficiencies and bending wave transmission across the junctions are also taken as spatially averaged values. In this way these metrics represent a point to point transmission between the rooms. However, for an isotropic large wall of uniform thickness the sound transmission coef cient of a plane wave depends on the incident angle θ between the direction of propagation of the incident plane wave and the normal to the plane of the wall. Below the critical frequency, the need to use a limiting angle is avoided by following the average diffuse eld single. This is possible by setting the coincidence angle θ_c equal to 90°, which does not exist below the critical frequency. A major advantage of this approach is that there is only a very slight discontinuity at the critical frequency as observed by [20]. As discussed before that the partition between the adjacent rooms may either be a homogeneous single wall element or consists of an assembly of components such as doors and portals as show in Figure **4.7**. Again, it is used an idea of segmenting the individual building elements into finite size patches and compute transmission coefficients based on incidence angle of the plane wave on the patches [80,82].

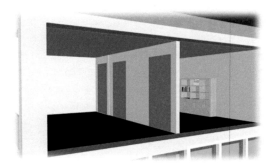

Figure 4.7: *An example of adjacent rooms separated by direct partition (i.e. the separating element) between source and receiving rooms*

In the example of a monolithic infinite plate structure, the angle dependent transmission coefficient $\tau(\theta)$ is given by Equation **3.31** in Chapter **3**. We now can calculate angle-dependent transmission coefficient which is a function of frequency angle-dependent radiation efficiency for a finite panel with rigid boundary conditions (such as windows, doors). Above the critical frequency, the transmission coefficient, for direct sound field, is calculated by using Equation **3.31** (with $\sigma(\theta) = \frac{1}{\cos\theta}$), whereas for diffuse field it is calculated by using Equation 20 (with $\sigma(\theta_c)$ form Equation **21**). Below the critical frequency, the sound transmission coef cient, for direct sound field, is calculated as the sum of Equation **3.35** and Equation **3.42** (with radiation efficiency from Equation **3.39** (with $g = \cos\theta$)) and for diffuse sound field it is calculated as the sum of Equation **3.35** and Equation **3.43** (with radiation efficiency from Equation **3.44**).

As another option, the angle dependent transmission coefficient for a small segment (i.e. a patch) on an infinite panel can be calculated by using spatial windowing technique introduced by Villot in [**24**]. In spatial windowing technique, the concept is to compute the radiated power where only a small area $S_{p,i}$ (of length i.e. L_x and width L_y) of the wall element i, contributes to the sound radiations. Hence, the radiation efficiency for a small segmented is computed and from there angle dependent transmission coefficient for finite size of patches on the walls can be predicted. It is shown in [**24**] that the results are closer to measurement results than other simple procedures. The next step is applying spatial windowing technique for small patches on the walls to the radiation process by correcting the transmission factor of infinite structure to obtain the corresponding transmission factor of the finite structure denoted as $\tau_p(\theta_p)$. As a result, we get the final transmission coefficient in case of single patch on the wall. The details for calculating radiation efficiencies $\sigma(k_0 \sin\theta_p)$ are described in [**24**].

$$\tau_p(\theta_p) = \tau(\theta)\big(\sigma(k_0 \sin\theta_p)\cos\theta_p\big) \tag{4.17}$$

4.2.5.2.   Flanking Sound Transmission

Flanking transmission is a more complex phenomenon than direct sound transmission as it involves building elements connected to each other through junctions. It is very important to take into account the bending waves while considering the flanking transmission. The bending wave travels through one element in the source room hits at the junction and travels to other element in the receiving room. The detailed theoretical concept for the calculation of sound power from flanking transmission is discussed in Chapter **3**. Once the transmission coefficients for individual patches are computed, it can be proceeded towards calculating transmission coefficients for each path ij from source room to the receive room defined in ISO [6] and is given in Equation **4.18**. Here S_i and S_j are the surface areas, and τ_i and τ_j are the transmission coefficients the flanking element i and j of the source and receiving rooms respectively. The surface area of the partition between the source and receiver rooms is denoted by S_D and vibration transmission over junction between the elements i and element j is represented by $d_{v,ij}$, which can be measured in accordance with ISO 10848-3 or ISO 10848-4.

$$\tau_{ij} = \sqrt{\tau_i}\sqrt{\tau_j} \cdot d_{v,ij} \cdot \frac{\sqrt{S_i S_j}}{S_D} \tag{4.18}$$

As we divide the individual wall elements into patches, the transmission coefficient of ij^{th} path from source room to the receiver room wall elements in terms of transmission coefficients of each patch of ith element of source room and each patch of jth element of the receiver room is given by Equation **4.19**.

$$\tau_{ij} = \sum_{all\,p}^{i} \sqrt{\frac{\tau_{p,i} \cdot S_{p,i}}{S_i}} \cdot \sum_{all\,p}^{j} \sqrt{\frac{\tau_{p,j} \cdot S_{p,j}}{S_j}} \cdot d_{v,ij} \cdot \frac{\sqrt{S_i S_j}}{S_D} \tag{4.19}$$

Here, $\tau_{p,i}$ and $\tau_{p,j}$ are the transmission coefficients and $S_{p,i}$ and $S_{p,j}$ are the surface areas of single patch p, on i^{th} and j^{th} elements of source and receiver rooms respectively.

4.2.6.   Sound Field in the Receiver Room

Once the sound is transmitted from source room through building elements via direct as well as flanking elements, it is radiated from the receiver room walls to the receiver end. From ISO [6], the sound power transmitted from i^{th} element of the source room to j^{th} element of the receiver room for direct and flanking paths is defined by Equation **4.20**, which is the final sound power of any radiating element j in the receiver room.

$$W_{R,ij} = \tau_{ij} \frac{S_D}{S_i} W_{s,i} \tag{4.20}$$

Using Equation **4.19** in Equation **4.20**, we get the expression of radiated sound power in the following form.

$$W_{R,ij} = W_{s,i} \frac{S_D}{S_i} \left[\frac{\sqrt{S_i S_j}}{S_D} \cdot d_{v,ij} \cdot \sum_{all\ p}^{i} \sqrt{\frac{\tau_{p,i} \cdot S_{p,i}}{S_i}} \sum_{all\ p}^{j} \sqrt{\frac{\tau_{p,j} \cdot S_{p,j}}{S_j}} \right] \qquad (4.21)$$

In Equation **4.20**, the sound power $W_{s,i}$ is obtained by taking sum of incident sound power of all the single patch from Equation **4.16** and using in Equation **4.21**, the expression of sound power for the receiver room can be written in the following form.

$$W_{R,ij} = \frac{W_a d_{v,ij} \sqrt{S_i S_j}}{S_i} \cdot \left(\sum_{all\ p}^{i} \left(\frac{F_{p,i}}{4\pi} + \frac{S_{p,i}}{A_s} \right) \sqrt{\frac{\tau_{p,i} S_{p,i}}{S_i}} \right) \cdot \left(\sum_{all\ p}^{j} \sqrt{\frac{\tau_{p,j} S_{p,j}}{S_j}} \right) \qquad (4.22)$$

In simplified approach (Section **4.1**), it assumed that the sound is apparently radiated from a single point representing the whole wave pattern on the wall elements of the receiver room. Therefore, the radiating elements (i.e. walls) in the receiving room were represented by single point sources. In this extended model, each radiating element j, of the receiver room is represented by a set of evenly distributed point sources (i.e. known as secondary sources) on its surface as shown in the Figure **4.8**.

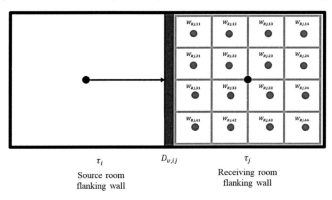

Figure 4.8: *Segmenting flanking wall of receiver room (SS for energy distribution)*

At this point, we can distribute the transmitted acoustic power $W_{R,ij}$, radiated by element j, among these secondary sources homogeneously by a factor $\frac{1}{P_j}$, where P_j are the total number of secondary sources (SS) on element j. The sound energy $W_{Rp,ij}$, radiated by a single secondary source of wall element j, with $Q_{Rp,j}$ as its directivity is then calculated from Equation **4.23**.

$$W_{Rp,ij} = \frac{W_a d_{v,ij}\sqrt{S_i S_j}}{S_i} \cdot \frac{1}{P_j}\left(\sum_{all\,p}^{i}\left(\frac{F_{p,i}}{4\pi} + \frac{S_{p,i}}{A_s}\right)\sqrt{\frac{\tau_{p,i}S_{p,i}}{S_i}}\right)\left(\sum_{all\,p}^{j}\sqrt{\frac{\tau_{p,j}S_{p,j}}{S_j}}\right) \tag{4.23}$$

The mean squared sound pressure of a secondary source for path ij in the receiving room can be derived from Equation **4.24**.

$$p_{Rp,ij}^2 = \rho_o c \cdot W_{Rp,ij} \cdot \left(\frac{Q_{Rp,j}}{4\pi r_{p,j}^2} + \frac{4}{A_R}\right) \tag{4.24}$$

Using $W_{Rp,ij}$ from Equation **4.23** into Equation **4.24** we get the final sound pressure for ij transmission path given in Equation **4.25**.

$$p_{Rp,ij}^2 = \frac{\rho_o c\, W_a\, d_{v,ij}}{S_i} \cdot \frac{1}{P_j}\left(\sum_{all\,p}^{i}\left(\sqrt{S_{p,i}\tau_{p,i}(\theta)}\frac{F_{p,i}}{4\pi} + \sqrt{S_{p,i}\tau_{p,i}}\frac{S_{p,i}}{A_s}\right)\right)$$
$$\cdot\left(\sum_{all\,p}^{j}\left(\sqrt{S_{p,j}\tau_{p,j}(\theta)}\frac{Q_{Rp,j}}{4\pi r_{p,j}^2} + \sqrt{S_{p,j}\tau_{p,j}}\cdot\frac{4}{A_R}\right)\right) \tag{4.25}$$

In Equation **2.25**, $r_{p,j}$ represents the distance between the acoustic centres of the radiating secondary source p of the wall element j of receiver room to evaluation point (i.e. position of the receiver). Finally, the time domain representation of the binaural signal at receiver point is obtained by introducing room impulse response of the receiver room and the HRIR filters for each secondary source to the receiver depending on its position and orientation relative to the secondary sources [80].

$$h_{R,ij}(t) = \sqrt{\frac{\rho_o c W_a d_{v,ij}}{S_i P_j}}\left[\sum_{all\,p}^{i}\sqrt[4]{\tau_{p,i}(\theta)S_{p,i}}\left(\sqrt{\frac{F_{p,i}}{4\pi}}\delta\left(t - \frac{r_{p,i}}{c}\right)\right.\right.$$
$$\left.+ \sqrt{\frac{S_{p,i}}{A_s}}h_{sp,i}(t)\right)\sum_{all\,p}^{j}\sqrt[4]{\tau_{p,j}(\theta)S_{p,j}}\left(\sqrt{\frac{Q_{Rp,j}}{4\pi r_{p,j}^2}}HRIR\left(t\right.\right.$$
$$\left.\left.\left.- \frac{r_{p,j}}{c}, \theta_{p,j}, \varphi_{p,j}\right) + \sqrt{\frac{4}{A_R}}h_{Rp,j}(t)\right)\right] \tag{4.26}$$

All $h(t)$ are statistically valid for all points inside both the source and the receiving rooms that is why $h(t)$ can be synthesized before implementing the auralization filter chain. However, as it can be assumed that $h(t)$ does not vary significantly for different source/receiver positions, hence, $h_{Rp,j}(t)$ and $h_{sp,i}(t)$ can be computed independently to avoid coherent interferences in the reverberant field coming from different radiating elements [80].

4.3. Façade Sound Insulation Filters: (Outdoor Scenes)

Sound transmission from an outdoor sound source, such as a vehicle, into a building is a complex process. Outdoor sound sources such as motorbikes, cars, buses or trucks are directional and may include strong low frequency sounds in their spectrum [22] and these sources also move. The direct sound propagation between the source and the façade creates the largest impact on the sound field in the receiving room. There might also exist multiple propagation paths from a vehicle to various parts of the building, for example, reflections from ground and surrounding buildings. Sound transmission into a building through façade is also complicated because of the varying angles of incidence of a vehicle noise, multiple transmission paths within the building and the fact that these phenomena all vary greatly with the frequency of the sound source [45]. This section of thesis introduces the procedures to estimate indoor sound levels because of outdoor transient noise sources (e.g. directivity) which means the façade sound insulation. The procedures to design sound transmission filters through façades should cover sound transmission losses of exterior walls, roof constructions and windows. Again, the method of segmenting the individual building elements into finite size of patches/segments, known as secondary sound sources (SS), is used as the exterior walls of common buildings are consist of an assembly of two or more parts or surfaces (e.g. windows etc.). In this respect, at first hand, we need to understand the basics of the outdoor sound propagation phenomenon, which we discus in next section in detail.

4.3.1. Outdoor Sound Propagation Model

The outdoor sound propagation models are important in understanding the perceptual effects of the sound fields of urban scenarios under building acoustical conditions. In previous studies [65,66] the façade sound insulation models are developed for outdoor sound sources which are based on the propagation of direct sound field originating from these sources. However, these models ignore important wave effects that contribute toward the perception of sound inside the buildings significantly. These effects are, generally, reflections from surrounding buildings and from the ground. Diffraction from edges or corners of the buildings can be usually ignored due to their low sound level in comparison with the direct and reflected sound. Different outdoor sound propagation models are combined with façade sound insulation model to provide a complete simulation of the outdoor scenes to achieve auralize of outdoor sound sources inside a dwelling. The outdoor sound propagation model used is partially based on the sound path algorithm that is implemented in virtual acoustics software (VA) developed at ITA, RWTH Aachen [**60,67**]. The VA software provides the opportunity to calculate the reflection and diffraction paths as well as the arbitrary combinations of both (i.e. reflections and diffractions up to a given order for certain building structures). These paths are calculated by different algorithms and their individual results are combined for the final result, as shortly described in the next section.

4.3.1.1.   Reflection Model

In VA, an implementation of the ISM (Image Source Method) is used for so-called sound-path-algorithm for detection of the reflection paths from source to the receiver with required number or orders [**67**]. Generating an image source would be costly in terms of computational effort since geometric models of the building used in a scene might contain hundreds of façade faces and many of the resulting paths might be invalid (i.e. not audible). Thus, the algorithm utilizes a propagation tree as illustrated given in Figure **4.9**.

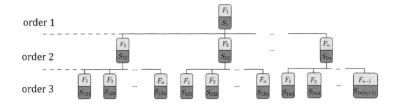

Figure 4.9: *Propagation tree of reflection path algorithm, where each node contains a face F and the corresponding image source S, adapted from*

The individual faces are inserted into the nodes N of the propagation tree. The root remains empty in the pre-processing as a place-holder for the source S. Each face $F \in \boldsymbol{F}$ is inserted as a child node $N_1 \in \{ N_{F_1}, N_{F_2}, \cdots, N_{F_n} \}$ of the root, where n is the total count of faces. On each of these faces the first reflection of the propagation path occurs, therefore, these nodes referred to as starting nodes. Higher order reflections are included by inserting each face $F_{i_2} \in F\{F_{i_1}\}$ into a node, where the node is bi-directionally linked to the parent node N_1 as one of its child nodes. This process is repeated until each starting node is connected to each corresponding child node. In general, $N_j = N_{F_{i_1}F_{i_2}\cdots F_{ij}}$ is connected to further children $N_{j+1} = N_{F_{i_1}F_{i_2}\cdots F_{ij+1}}$ until the depth j of the tree reaches the maximum reflection order m as illustrated in Figure **4.9**. Each node represents the basis of propagation path candidate, with the chain of faces as $F_{i_1} \rightarrow F_{i_2} \rightarrow \cdots \rightarrow F_{i_j}$. As a next step the source S is inserted into the root node of the propagation tree, and the image source are generated in accordance with the ISM. Initially, all non-illuminable starting nodes are neglected by using back-face culling between source node and the initial face of the nodes. The detailed explanation of the back-face culling can be found in [**68**]. For the remaining nodes first order image source S_{i_1} is created by mirroring the source S along face F_{i_1} and its position is saved in the corresponding node. Higher order image sources are set up through mirroring the image sources $S_{i_1 i_2 \cdots i_{j-1}}$ at face F_{i_j} with j being the order of the generated mirrored image sources $S_{i_1 i_2 \cdots i_j}$. This procedure is repeated for all non-neglected nodes of the tree until each node N_j can be used to build a chain of j image sources. After the creation of image sources, each node represents a propagation path candidate together with S and R. Construction of paths is done by back-tracking R along an image source chain represented by N to S. The position of R is saved as the starting point of the propagation path, denoted by P_{start} and the image source position $S_{i_1 i_2 \cdots i_j}$ is saved as target point P_{target} respectively. Between P_{start} and P_{target}, a line segment is built. If the line segment intersects with the corresponding face F of the image source, the intersection point I_F is registered as a valid reflection point. Otherwise, whole path is considered invalid and will be neglected. After the registration of a valid reflection point, I_F is set as the new P_{start} and the calculation of reflection points is repeated until the root node is reached. Finally, the source S is pushed to the front of the propagation path chain. A chain of ith order is given as $S \rightarrow S_{n_1} \rightarrow S_{n_1 n_2} \rightarrow \cdots \rightarrow S_{n_1 n_2 \cdots n_{i-1}} \rightarrow S_{n_1 n_2 \cdots n_i} \rightarrow R$, with source S, receiver R and image sources $S_{n_1 \cdots n_i}$. The indices n_k with $n_k \neq n_{k\pm 1}$ denote the number of the reflected walls. The algorithm returns the interaction points and moves on to the next list. After repeating the construction and repeating phase for all paths, all reflection paths up to a given order are deterministically determined. This model sets the basis for the construction of the impulse responses between the outdoor source and the façade patches [**67**].

4.3.2. Filter Design

ISO [17] provides basic guidelines for the airborne sound insulation against outdoor sound sources. In this ISO standard, it is made an assumption that the outdoor sound field is diffuse. Hence, the transmission factors (sound insulation) are predicted by considering a source position at 45 degrees angle relative to the façade. However, in real outdoor scenes the sound source might be present at specific location relative to the façade. Therefore, we take into account the direct part and early reflection part of the sound field hitting at surfaces of the exposed building façades at their specific angles of incidences. Hence, it is considered angle dependent radiation efficiency $\sigma(\theta)$ to get angle dependent transmission coefficients. The angle dependent transmission factors (given in Equation **4.20**) are used for each patch (i.e. façade component) for direct sound field and diffuse transmission factor (given in Equation **3.26**) are used for the early reflection and diffuse part of the sound field for the filter design process of sound insulation. Nevertheless, the direct sound transmission paths (Dd) for each small segment of façade elements (i.e. secondary sources) are only taken into account as it is assumed that the sound transmission from each secondary source is independent of the sound transmission from others [**22**].

Figure 4.10: *Exterior walls of the receiving room (The glass windows are secondary sources)*

In order to calculate the insulation filters, in first place, the source directivities and the outdoor sound propagation model are used for computing sound pressure at the surfaces of the façade patches. The outdoor sound field is taken as a combination of direct sound and early reflections. The early reflections of the sound field are obtained as an outcome of outdoor sound propagation model as discussed in the previous section. Secondly, the receiving room acoustics is implemented that includes

78

the receiving room reverberation based on room geometry, absorption and transfer functions between secondary sources and the receiver. As mentioned before, the façade of buildings may be a single homogeneous element or it may consist of an assembly two or more parts as show in Figure **4.10**. Once we get the corresponding transmission factors for each patch of façades, sound insulation filters are designed based on presented model in Section **4.2** with additional upgradations discussed below.

Let us assume an outdoor source with directivity Q_s. From Equation **4.12**, the mean squared sound pressure in energetic notations at any point on the external surfaces of the façade at a distance r from the source is derived as given by Equation **4.27**. However, there is no source room involved rather the sound source is outdoor moving vehicle, therefore, second part in brackets in Equation **4.12** is omitted for outdoor cases. Nevertheless, additional terms must be introduced in Equation **4.27**, which are the energy contributions of early reflections and diffractions from the surrounding building façades and edges respectively.

$$\tilde{p}_s^2 = \rho_o c W_a \left(\frac{Q_{s,p}}{4\pi r_{s,p}^2} \right) \tag{4.27}$$

Hence, the sound power at any point on the façade can be calculated with a simple modification to the stationary sound field in the ordinary room that is to take into account the direct sound field and the early reflection part of the sound field. Under free field conditions the direct incident sound power on a secondary sound source p with a surface area of $S_{s,p}$, denoted by $W_{s,p}$ is given by Equation **4.39**, where $r_{s,p}$ is the distance from the source to the infinitesimal element $dS_{s,p}$ on the façade secondary source and $\theta_{s,p}$ is the incidence angle of the plan wave. Thus the incident power on each secondary source of an element is calculated as,

$$W_{s,p} = W_a \int_{S_{s,p}} \frac{Q_{s,p}}{4\pi r_{s,p}^2} \left| \cos \theta_{s,p} \right| dS_{s,p} \tag{4.28}$$

The integral in Equation **4.28** can be written as $F_{s,p} = \int_{S_{s,p}} \frac{Q_{s,p}}{4\pi r_{s,p}^2} \left| \cos \theta_{s,p} \right| dS_{s,p}$ by assuming that Q_s, $r_{s,p}$ and $\theta_{s,p}$ do not vary significantly along $S_{s,p}$ it can be approximated as $F_{s,p} \approx \frac{S_{s,p} \cdot Q_s}{4\pi r_{s,p}^2} \left| \cos \theta_{s,p} \right|$. Using Equation **4.39** and the transmission coefficients $\tau_{s,p}(\theta_{s,p})$ from Equation **4.17** (which is angle dependent transmission coefficient) and τ_d from Equation **3.9** (which is the transmission coefficient under diffuse sound field) for the direct sound field and the early reflection part respectively, the sound power transmitted from secondary source to receiver room by one secondary source is given by Equation **4.29**.

$$W_{r,p} = W_{s,p} \frac{S_{s,p}}{S_i} = W_a \frac{S_{s,p}}{S_i} \cdot \left(F_{s,p} \tau_{s,p}(\theta_{s,p}) \right) \tag{4.29}$$

Finally the contribution of the **Dd** path for the a single secondary source to the mean squared pressure in the receiving room is derived by using the expression given by Equation **4.30**, with $Q_{r,p}$ as directivity of secondary source and A_R, as the equivalent absorption area of the room.

$$p^2_{r,sp} = \rho_0 c \cdot W_{r,p} \left(\frac{Q_{r,p}}{4\pi r^2_{r,p}} + \frac{4}{A_R} \right) \tag{4.30}$$

Inserting Equation **4.29** in Equation **4.30**, the sound pressure for single '**SS**' is given by Equation **4.31**, where S_i is the area of the walls and $r_{r,p}$ is the distance of receiver from secondary source.

$$p^2_{R,sp} = \frac{\rho_0 c W_a \cdot \tau_p(\theta_{s,p}) \cdot S_s}{S_i} \left(\frac{Q_{r,p} \cdot F_p}{16\pi^2 r^2_{r,p}} + \frac{F_p}{\pi A_R} \right) \tag{4.31}$$

In time domain, the corresponding impulse response included and the transfer function $\tau_p(\theta_{s,p})$ and τ_d, and $h_R(t)$ as the normalized reverberation tail of the receiver room. The final result in time domain from source to receiver for one secondary source as radiating element is given by Equation **4.32**. [80]

$$h_{R,ss}(t) = \sqrt{\frac{\rho_0 c \cdot W_a S_s}{S_i}} \sum_{all\ p}^{j} \left[\sqrt{\tau_p(\theta_{s,p})} \left(\sqrt{\frac{Q_{j,p} \cdot F_p}{16\pi^2 r^2_{r,p}}} HRIR \left(t - \frac{r_{s,p} + r_{r,p}}{c} \right) \right. \right.$$
$$\left. \left. + \sqrt{\frac{F_p}{\pi A_R}} h_{Rp,j} \left(t - \frac{r_{s,p}}{c} \right) \right) \right] \tag{4.32}$$

4.4. Filter Rendering

In previous sections it is discussed the airborne sound insulation and impulse response filters construction for indoor (adjacent rooms) and outdoor (facade) environments based on ISO [6,17] and available research with the improvements. Subsequently, these filters are applied as the input data for auralization of sound insulation for different case studies presented in Chapter 5. In this section, the filters cascading and rendering methods are described in two parts. At first, filters rendering for outdoor environments is discussed with the description of a complete rendering chain in the form of flow charts, given in annex A. For outdoor urban environments, the sound propagation model returns a set of paths for the direct sound field, early reflections independent of each other. These paths are required to be combined in the form of an impulse response to get a complete response of the outdoor scene at facades of a building. Two methods used to combine both direct and reflection paths which are calculated in the previous sections. These methods are named as 1) "sorted-sequence path algorithm" and 2) "any sequence path algorithm" [67]. The calculation steps are the same as described in the previous sections.

4.5. Auralization

Once, the filters for sound insulation are calculated as described in the previous sections, auralization can be performed. Auralization makes sound pressure in the receiving room audible at the ears of the listener by replay of the signals by an appropriate audio reproduction equipment. In this section, the methods for binaural reproduction of the sound insulation filter are very briefly discussed. It is not the scope of the thesis to explain the technology behind auralization processes, such as head related transfer function (HRTFs) from dummy head recordings, convolution techniques, and the basics of sound reproduction and dynamics. We only discuss the important aspects that are used for representing binaural signals with perspective to sound insulation.

4.5.1. Source Signals

Sound pressure signals of sources are the primary data characterizing the source. They must be obtained in a free field, thus giving a unique feature of the source without any impact from the environment. Physical modelling can lead to source signals [76]. Stationary signals of compact sources can be recorded in anechoic chambers known as "dry signal". A large variety of dry signals are available in databases [**77**]. On the other hand, signals from the moving sound sources can be measured and classified in terms of sound characteristics and then synthesized [**73**].

4.5.2. Interpolation

From an input time signal $s(t)$ and transfer function, the time signal at the output of any LTI system can be calculated by means of convolution techniques. However, the filter length and its digital resolution are important parameters for the convolution technique. As mentioned in Chapter **3**, the audio frequency range for human hearing is typically 20Hz to 20KHz, nonetheless, we consider building acoustics frequency range to be defined by one-third-octave-bands from 50Hz to 5KHz. For signal processing, these quantities have to be turned into frequency spectra with a practical number of frequency bins. In case of sound insulation filters, an input with **21** values in one-third octave bands is given (ranging 50Hz to 5KHz). To obtain a frequency spectrum with **4097** lines, these values have to be interpolated. This is done applying cubic spline interpolation. Also extrapolation is necessary which is particularly crucial for low frequencies.

4.5.3. Binaural Techniques

In the receiving room, the building elements are excited by structural waves due to the sound transmission from the source room.

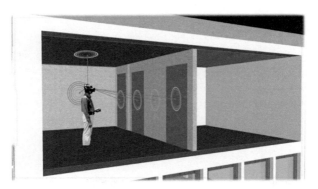

Figure 4.11: *Binaural reproduction from secondary sources located at different position. HRTF are applied for each secondary source*

They act as secondary sound sources located at different positions and orientations relative to the listener's ears. Therefore, they are required to perceptually localize to create a spatial impression of listening in the room. Hence, it is necessary to consider an auralization with measured or individualized binaural signals by head related impulse responses (HRIR) or corresponding head related transfer function (HRTF). In the auralization process, the HRTFs of the ITA dummy head are used. Therefore, to experience the impression of the receiving room, impulse responses are simulated for the receiving room as described in detail in the previous section for each pair of secondary source and listener positions.

4.5.4. Signal Presentation for Listening

For presentation of signals resulting from building-acoustic auralization, calibration of the absolute loudness is required. When listening to classical room-acoustical auralization, the colouration, the spaciousness or lateral fraction are important which do not change much with level. In contrast, in building-acoustical auralization, the level is the most important aspect, together with colouration. Therefore, care has to be taken for the reproduction of both the correct absolute level and the relative level between source and receiving room. Since some room-to-room situations have level differences of 50dB and more, care has to be taken for not wasting valuable signal-to-noise ratio in the signal chain. If the absolute level of the sound signals is to be reproduced, a calibration of the replay chain has to be done.

4.5.5. Headphone Equalization

For presentation of binaural signals, mostly headphones are used since they are easy to handle and provide the necessary separation of the left and right ear signal. It is, however, problematic to use different types of headphones since they have different transfer functions. Ideally, the headphone transfer functions have to be equalized for reproduction with reference to free field (free-field equalized headphone).

5

Implementation and Verification

This chapter, on one hand, implements the filters for auralization and on the other hand, describes the accuracy of the extended airborne sound insulation model, the quality of corresponding filters, and verification of filters' auralization for both selected indoor and outdoor environments. Three crucial parts are recognised as the areas for the verification and validation of auralization chain. First part is the extended sound insulation model itself that takes into account the limitations of the previous models. The predicted results of sound insulation metrics from the extended model are compared with those of ISO standards (ISO 12354-1 [6], ISO 12354-3 [17]) and with the measurements of some building acoustic structures and conditions in terms of performance of the building elements. The most important sound insulation metric, which is the standardized sound level difference D_{nT} between source and the receiving rooms, is verified by reproducing it by tuning according to the standard conditions of ISO [6,17]. As a second step for verification of extended model, the results of sound transmission through structural elements of the adjacent rooms as well as for façades are compared with the traditional approaches, such as, diffuse field approach under ISO standard settings.

All parts are assessed in a simple and transparent way, while presenting data in plots, graphs and building sketches. Thus, the focus is more on the verifications of the overall concept as well as showing that the results match with available standards under same standard conditions. It is also shown that the extended sound insulation model might not only be used as prediction tool for sound insulation metrics for building elements but also provide the opportunity to construct filters for auralization for building acoustical conditions differing from standard settings (provided by ISO [6,17]). The first step towards verification is getting started with the implementation of the sound insulation models in virtual building acoustics

(VBA) framework [61] by taking different indoor and outdoor environment scenes as case studies, which are described in the following sections.

5.1. Built Environments (Case Studies)

The implementation is carried out by selecting two main built environments. The first environment is an indoor environment with façade, such as, a residential apartment, an office building or a worksite as a real-world case study while the second one is an outdoor environment that consists of two urban scenes of different geometrical configurations and complexities. The VBA framework implements sound insulation filters for evaluation of the performance of individual building elements as well as the overall performance of the built environments. Verification of sound insulation filters is carried out for both case studies with reference to standard settings (proposed in ISO [6,17]). For the indoor scene, the purpose is to evaluate the results of predicted airborne sound insulation metrics, such as sound level differences, from the developed sound insulation model for the adjacent offices where one office is taken as source room, and the other room is taken as receiving room. For outdoor scenes, the evaluation of filters is carried out for building façade elements that are exposed to an outdoor sound sources. At first, façade sound insulation filters are verified and later on implemented for urban scenes along with the implementation of outdoor sound propagation model.

a   b

Figure 5.1: *ITA-Building – Selected work site for verification of airborne sound insulation model (VBA Framework) (a) Virtual Building, (b) Real Building*

The building of the Institute of Technical Acoustics (ITA) RWTH Aachen University, Germany is selected as a real scene office premise for which the sound insulation metrics are simulated using VBA. The construction design of the ITA building allows us to select different adjacent offices as source and receiving rooms for simulating different indoor scenes. ITA building is a triple storey building with a basement, a ground floor and first floor. The 3D graphical view of the real and virtual ITA building are shown in Figure **5.1** and its corresponding construction map in Figure **5.2**.

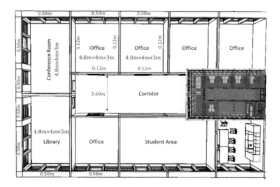

Figure 5.2: *ITA-Building ground floor plan*

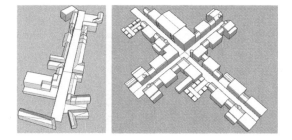

Figure 5.3: *Geometrical models for urban scenes: (left side) is simple street canyon and (right side) is a crossroad junction square*

The implementation of VBA for outdoor environments is performed by designing two urban street scenes with different complexities of their geometrical models which are; 1) a straight street canyon and 2) a crossroad junction street. These complexities are important in order to explore different aspects of sound propagations such as, reflections from the surrounding building geometries and the position of the source on the road relative to the facades etc., and the influence of sound transmission through complex building façades. Figure **5.3** shows two selected models where the first model is simple street canyon with a straight road along with dwellings on both sides of it and the second one is a crossroad junction. Different building with different complex structures are designed alongside the roads with different heights, widths and roof types in order to make the outdoor sound propagation model more realistic and comparable to real-world environments.

5.2. Evaluation for Adjacent Rooms (Indoor Case)

An appropriate and simple way for the verification of the extended sound insulation model is to start from analysing important insulation metrics of different elements of the selected dwellings such as, transmission coefficients. The results of transmission coefficients (or sound reduction indices) from the developed model are verified by comparing with that of ISO [**6**] and available measured data. The reproducibility of the level difference D_{nT} between source and receiver rooms is the main aspect for the verification. The purpose of this comparison is primarily to validate the extended approach in compliance with the standards (i.e. ISO). Two adjacent identical office are selected in ITA building as source and receiving rooms both with dimensions $4m \times 5m \times 3m$, as case study, but not with the intention to exactly model this building with the aim to compare with exact in-situ measurements. Only the exact geometrical dimension of the ITA building rooms are taken into account, however the material data is not exact. The building construction data and material properties of these rooms are given in Table **5.1**, which we used as input data for calculation of sound insulation metrics. The verification is based on comparison between filters output and ISO [**6**] input data. The reverberation times are simulated for both source and receiving room which are $0.7\,s$ each room at $500Hz$, and is used as input data for auralization.

Table 5.1: *The material/geometry properties of source and receiver room*

	Dimensions $(L \times W)\,m$	Thickness m	Material	c_L	η_{int}	f_c Hz
Main Partition	5×3	0.12	Concrete	3800	0.005	140.8
Flanking Walls	4×3	0.12	Concrete	3800	0.005	140.8
Ceiling/Floor	4×5	0.3	Concrete	3800	0.005	56.33

5.2.1. Verification of Level Difference (D_{nT})

The main partition between the selected rooms is 5×3 meters concrete wall with thickness of $120\ mm$, mass per unit surface area $246\frac{kg}{m^2}$ and the internal loss factor is 0.005. This partition is an assembly of doors and concrete, therefore for direct sound transmission the partition is segmented into patches (i.e. segments, see Chapter 4 for details) as shown in Figure 5.4.

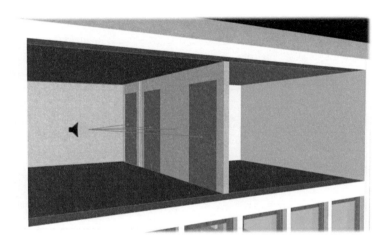

Figure 5.4: *Adjacent source-receiver rooms with main partition divided into patches (i.e. small segments)*

For each patch we calculate the transmission coefficients using extended insulation model. The angle dependent transmission factors $\tau_p(\theta_p)$ for each patch are calculated using Equation 4.20 (Chapter 4), whereas, for diffuse field, the transmission coefficients τ_d are calculated using Equation 3.27 and Equation 3.33. To reproduce the results from extended approach, the D_{nT} values are obtained from the simulated sound pressure values for three different source and receiver positions including the normalization to the reverberation time of the receiving room. This can be interpreted as "virtual measurement" following the standard settings of ISO [6].

Figure 5.5 shows the computed D_{nT} values in one-third octave band which are averaged over three random source positions and three random receiver positions

[69]. In the same gure, the predicted D_{nT} results following ISO [6] (i.e. based on transmission coefficients τ_{diff}) are compared with that of the extended approach D_{nT} results. The differences between D_{nT} values of both ISO [6] and extended approach are also shown. From the Figure 5.5, it can be seen that the extended model results are in good agreement with that computed from the ISO standard (i.e. diffuse field approximations) in the case of adjacent rooms. The maximum difference between D_{nT} values of both approaches is below 1.9 dB.

Furthermore, in Figure 5.6 and Figure 5.7, it is compared sound insulation of non-standard settings in real building situation. This (i.e. non-standard settings) can be any condition outside the prerequisites for the definition of sound insulation, such as 1.5m distance between source and receiver positions and room boundaries or source directivities. For this we take two cases as non-standard settings (i.e. source/receiver configurations), where the source is modelled as a HiFi stereo sound system with loudspeakers' directivities pointing to the centre of the source room. In first configuration, the system is placed 0.3 m away from one of the flanking wall and in the second configuration it is placed 0.3 m away from the main partition pointing towards centre of the room.

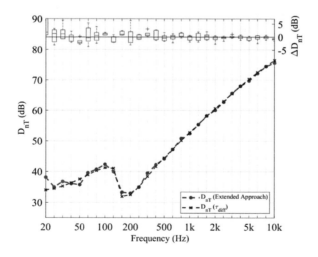

Figure 5.5: *Comparison of standardized level differences, calculated from extended model (D_{nT}) and ISO ($D_{nT}(\tau_{diff})$) between source and receiving rooms based on standard settings of ISO (for both models)*

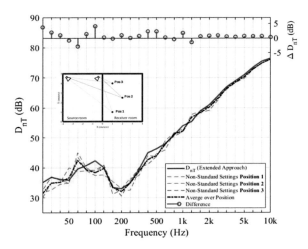

Figure 5.6: *Case-I*: *Comparisons of standardized level differences D_{nT} between non-standard configurations for three receiver positions and their average (black) and standard configurations (red), both computed from extended model*

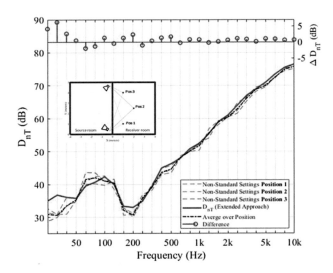

Figure 5.7: *Case-II*: *Comparisons of standardized level differences D_{nT} between non-standard configurations for three receiver positions and their average (black) and standard configurations (red), both computed from extended model*

5.2.2. Comparison with Measurements

Measurements were carried out to compare the simulated results (i.e. from extended model) with the actual results for both standard and non-standard settings for an adjacent rooms of similar dimensions. As mentioned above, material properties of building elements of the actual rooms may not be exact. For this reason, we can only compare differences between settings of sources and receivers but not the absolute values of simulation vs. measurement. The main partition separating the room has thickness of **120 mm**. This partition is plaster board with additional layers. There is no information about the actual material of the main partition (dry wall construction). The flanking walls, ceiling and floor are made of concrete. Figure **5.8** shows the schematic diagram of the adjacent rooms and the measurement procedure for both standard and non-standard settings.

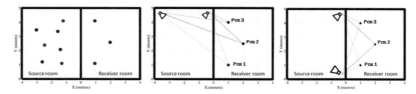

Figure 5.8: *Adjacent rooms: (**left**) Standard settings of source, (**middle** and **right**) non-standard settings of source (similar settings as in Figure **5.5**, Figure **5.6** and Figure **5.7**). Red dots are source and blue dots are receiver positions*

The measurements of the level differences were carried out according to ISO 10140. Four microphones (KE-4 Microphones) positioned in each room and three loudspeakers (ITA dodecahedron) positions in source room were measured. Sweep signals were used to measure the sound insulation. The details of the measurements are described in [**3**]. The reverberation times are derived from the room impulse responses for the standard and non-standard settings.

5.2.2.1.   Level Differences

The measured level differences $\Delta L = L_s - L_R$ and D_{nT} for the three settings are shown in Figure **5.9**, Figure **5.10** and Figure **5.11**. The measurements were repeated three times. Here, L_s is the average sound pressure level (dB) in source room (for three positions of source and four positions of microphones and L_R is the average sound pressure level (dB) in receiver room for three positions of microphones.

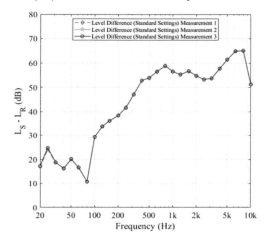

Figure 5.9: *Level difference between source and receiver rooms for standard settings: (Three measurements each with average over three receiver positions)*

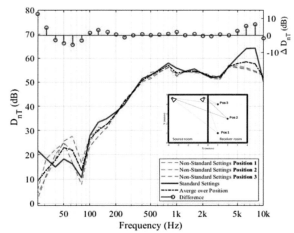

Figure 5.10: *Case-I: Comparison of measured D_{nT} between non-standard settings (three position and their average (black)) and standard settings (red). Difference in D_{nT} between averaged non-standard settings (black) and standard setting (red) (at top of figure)*

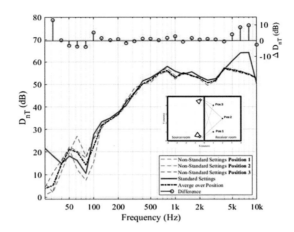

Figure 5.11: *Case-II: Comparison of measured D_{nT} between non-standard
settings (three position and their average (black)) and standard settings (red).
Difference in D_{nT} between averaged non-standard settings (black) and standard
setting (red) (at top of figure)*

Figure **5.9** shows the level difference of three measurements under standard
settings (according to Figure **5.8(left)**), which are averaged over three receiver
positions. Whereas, Figure **5.10** and Figure **5.11** compares the measured level
differences D_{nT} at three positions under non-standard settings (according to Figure
5.8 (middle, right)) with that of standard settings D_{nT} of Figure **5.9**.

From simulated (Figure **5.6** and Figure **5.7**) and measured (Figure **5.10** and
Figure **5.11**) results, similar trends are observed in D_{nT} curves for non-standard
configurations. The differences between standard-setting D_{nT} and non-standard
settings D_{nT} (averaged over three positions) are calculated for both simulations and
measurements as show in Figure **5.12** and Figure **5.13** respectively. It is observed
(from Figure **5.12** and Figure **5.13**) that the differences (ΔD_{nT}), between D_{nT} of
standard and non-standard settings for both simulations and measurements have
similar trend at mid frequency range in both cases (Case-I and Case-II). Below 150
Hz, no conclusion can be drawn since the simulation model cannot compute modal
effects. Above 5 kHz, above the critical frequency the measurement results may
include effects which are not covered by the simulation. The measured sound
insulation in the standard setting is much higher than in the non-standard setting.

Due to the very close source position near to the walls, specific bending wave excitation may lead to efficient sound transmission compared to the standard setting with its diffuse mix of incidence angles. Below the critical frequency, these results may lead to the conclusion that the level differences depend on source directivity, position and angle of incidence on the walls. The simulation and the measurement follow the same trend.

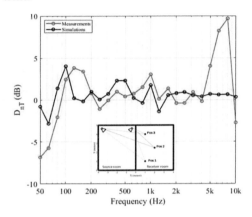

Figure 5.12: *Case-I: Difference (Δ) between D_{nT} of standard settings and D_{nT} of non-standard settings for simulations (blue) and for measurements (green)*

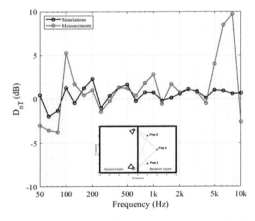

Figure 5.13: *Case-II: Difference (Δ) between D_{nT} of standard settings and D_{nT} of non-standard settings for simulations (blue) and for measurements (green)*

5.2.3. Visualization of Sound Fields

To further compare the results of the extended model with the classical sound insulation models (i.e. ISO standard), the sound pressure levels are computed and visualized for three different positions and orientations as shown in Figure **5.14** to Figure **5.16**.

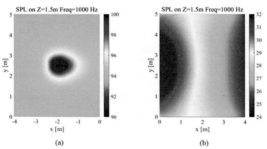

Figure 5.14: *Sound pressure level for **source position 1** from extended approach in (a) source room and (b) receiving room; (at z = 1.5 m high plane)*

The main purpose of sound field visualization is to see the effects of source directivity on the energy transmissions through different transmission paths. It is selected a plane at 1.5 m above the floor inside the source and the receiver rooms. The final room impulse responses, based on extended model, is calculated at receiver points by taking into account the direct and the reverberation sound fields of the source as well as the receiving rooms.

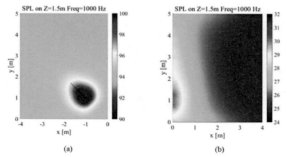

Figure 5.15: *Sound pressure level for **source position 2** from extended approach in (a) source room and (b) receiving room; (at z = 1.5 m high plane)*

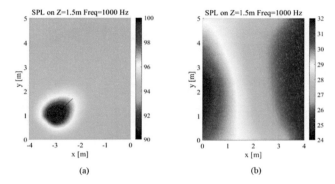

Figure 5.16: *Sound pressure level for **source position 3** from extended approach in (a) source room and (b) receiving room; (at z = 1.5 m high plane)*

Figure **5.14**, Figure **5.15** and Figure **5.16** describe the spatial variation of the sound fields inside the source room and receiving rooms for three positions and orientations of the sound source.

5.3. Verification of Façade Sound Insulation

The same concept is applied for the verification of façade sound insulation as discussed in the previous section for adjacent offices. However, diffuse field assumptions are not taken into account as there is no diffuse sound field outside the dwellings in real world.

Therefore, the reverberation part of the outdoor sound field does not exist, at least not if only the direct excitation path is considered. Nevertheless, the receiver room reverberation is taken into account. As an example, a corner office is selected at ITA building as receiving room with two facades facing as external wall elements of the receiver room. Figure **5.17** shows the building facades of the receiving room and the construction plan for the receiver room. In this case the sound source is an outdoor vehicle with specific directivity. The reproducibility of the level difference D_{nT} between outdoor to indoor is main aspect for verification.

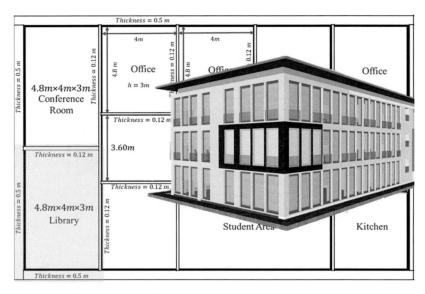

Figure 5.17: *ITA-Building – for façade sound insulation*

5.3.1. Verification of Level Difference (D_{nT})

In Figure **5.18**, the comparison of D_{nT} values of the extended approach and the diffuse field approximation (ISO) is presented for façade sound insulation of an office room with dimensions of $6.5 \times 4 \times 3 \, m^3$. The selected external walls (i.e. façades) of this office are an assembly of different materials and consists of single-pane glass windows connected through concrete pillars as shown in Figure **5.1**. Note that the real ITA building has double glazing but due to lack of information of the actual sound reduction index of the built-in windows, a hypothetic glass window is used. The height and width of each glass window are $2.5 \, m$ and $1 \, m$ respectively. The glass thickness is $8 \, mm$, density is $2500 \frac{kg}{m^3}$, and the internal loss factor is 0.004. Each window is modelled as a secondary sound source for the receiving and the sound insulation for each secondary source is computed independently as finite segments. In this way, the façade acts as multiple secondary sources which radiate sound energy to the receiver room. The transmission coefficients for each secondary source. It is assumed that the sound transmission of each secondary source is independent from the sound transmissions of others and have no interaction with each other in terms of transmission of bending waves across them.

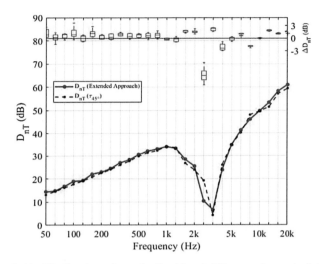

Figure 5.18: *Verification of standardized level difference, D_{nT} calculated from extended approach and ISO (both calculated for standard settings of source and receiver according to ISO)*

The procedure to obtain the D_{nT} values as a "virtual measurement" at exactly 45° incident angle on the façade, the condition of a plane wave incident is fulfilled by placing source at a very large distance (for example $500\ m$) to the façade and the outdoor sound level is obtained at the surface of façade. According to ISO [17], an addition $3\ dB$ is added to include the contribution of reflected energy from the façade. The indoor sound level from the extended approach is calculated by taking the average sound pressure level for three random positions in the receiving room from all secondary sources (i.e. windows). The reference outdoor $D_{nT,45°}$ referred to in the blue curve of Figure **5.18** is calculated based on [17]. The differences between the extended approach D_{nT} values and the ISO standard D_{nT} values are also shown in Figure **5.18**. Once the D_{nT} values from the extended model are verified and are found in good agreement with that of ISO under same standard setting, three example cases with non-standard settings (i.e. configurations) are created for outdoor sound source to evaluate the effects of source position in front of the façade. The evaluation is done to see how big the changes in the effective D_{nT} values of the receiving room are in case of moving outdoor sources.

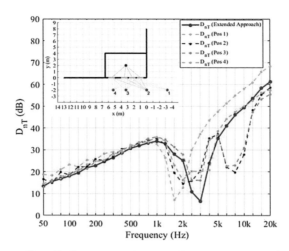

Figure 5.19: *Case I: Comparisons between the D_{nT}, calculated for four source positions at non-standard settings and the D_{nT} calculated for standard settings (red) (with omnidirectional source)*

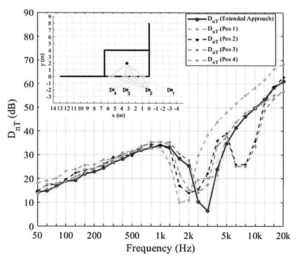

Figure 5.20: *Case II: Comparisons between the D_{nT}, calculated for four source positions at non-standard settings and the D_{nT} calculated for standard settings (red) (with directional source facing 90^0 away from façade)*

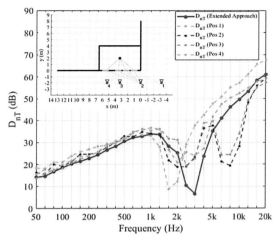

Figure 5.21: *Case III: Comparisons between the D_{nT}, calculated for four source positions at non-standard settings and the D_{nT} calculated for standard settings (red) (with directional source facing towards façade)*

Figure **5.19** compares the results of standard settings (red) with the results of such nonstandard configurations where the source is placed at four different positions such as the sound energy from source arrives at façade from different angles of incidences. The variations in the D_{nT} curves for the four positions of the sound source are because of the deviation from the standard situation concerning distance and angle of incidence of sound waves to different secondary sources (i.e. windows), hence, the actual sound insulation curves (which are from non-standard setting of ISO) differ from the standard results (which are standard setting of ISO). Similarly, Figure **5.20** and **5.21** compare the results of standard (red) and nonstandard settings but with different source directivities. At low frequencies, the main factor causing a curve shift is the incidence angle and the normal component of the particle velocity in relation to the mass law. At high frequencies, the coincidence effect of glass in relation to the incidence angle shows quite large differences as expected [80].

5.3.2. Visualization of Sound Fields (Outdoor)

Finally, the sound pressure level distribution in the receiving room for excitation from outdoor sources is visualized. Three positions and orientations of the sound source are selected. As expected, the outdoor sources have more influence on variation of angle-dependent incident sound power on the building elements and in consequence, on the different amount of energy transmitted through direct path. Figure **5.22** shows the spatial variation of the sound field inside the receiving room due to façade sound insulation. Here, we take into account the exterior walls of the receiving room as independent secondary sources (patches) consisting of glass windows of **8 mm** thickness.

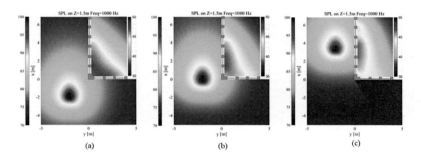

Figure 5.22: *Sound pressure level of outdoor source from façade to receiving room at z = 1.5 m, height from floor for three source positions in **a**, **b** and **c***

5.4. Extension to Urban Environments (Outdoor)

The extension of the urban environments is carried out by selecting two real world scenes, as mentioned in Section **5.1**. The first scene is a straight street canyon whereas the second scene includes a crossroad junction as shown in the Figure **5.23**. In the first scene, the sound source can only vary on a straight road, whereas in the second scene there can be many positions where the sound source can be. In first scene, the source is always in the direct line of sight of the façade of the building under evaluation. In other words, there is no road junction in this scene rather a straight street canyon and the sound sources are only on straight line in front of the façade as show in Figure **5.23(a)**. On the other hand, in the second scene, which is similar to the first scene with respect to the complexity of the building geometries. Here, however, a crossroad junction is introduced while the building under evaluation is situated at the corner of the junction with a corner room which is selected as receiving room at first floor of the building. In this way, two corner facades of the receiving room are facing towards the crossroad junction and are likely to be exposed to outdoor sound field depending on the routes of sources. Different routes are possible for second scene, as can be seen from Figure **5.23(b)**. The façades of both buildings (in both scenes) are excited by the direct sound of the source at different positions depending on the line of sight of the source to the facades. For example, when the source is placed at the crossroad junction in scene 2, the sound field excites both façades. Afterwards, when the source is moving away from the corner towards any predefined position, one of the façades loses the line of sight of the direct sound path. The sound pressure level differences are computed by following the same procedure at discussed in previous section and the results are discussed in the next sections.

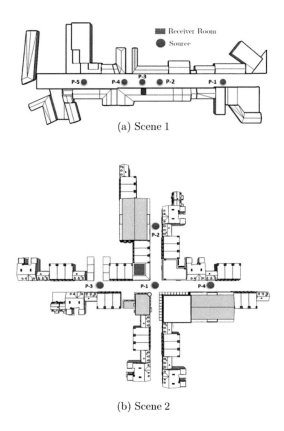

(a) Scene 1

(b) Scene 2

Figure 5.23: *Two urban environments: The buildings are marked as blue (a) Scene 1 - A street canyon (b) Scene 2 - A crossroad street junction*

5.4.1. Verification of Level Difference (D_{nT})

Similar procedures are followed for verification D_{nT} for the urban scenes 1 and scene 2. Figure **5.24** shows a comparison of D_{nT} values calculated based on complete outdoor sound propagation model and diffuse field approximation (ISO) of a receiving room in scene 1 with dimensions of $6.5 \times 4 \times 3\ m^3$.

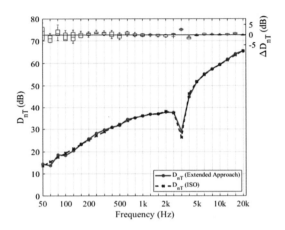

Figure 5.24: *Verification of standardized level difference, D_{nT} calculated from extended approach and ISO between outdoor source and receiving room based on standard settings of source and receiver*

The external walls (i.e. façades) of this receiving room are an assembly of different materials and consists of glass windows concrete walls. The height of each glass window and concrete wall is $2.5\ m$ whereas the width of each element is $1\ m$. The glass windows are single glazed with thickness $8\ mm$. The density is $2500 \frac{kg}{m^3}$ and the internal loss factor is 0.001. As each component of the façade acts as a secondary source (SS) for receiving room therefore sound insulation curve for each SS is computed independently considering them as finite segments (patches). Further in Figure **5.24** the differences ΔD_{nT} in dB, are computed from diffuse field approximation in ISO [17] and from the extended model with the standard setting (proposed in ISO standards [17]). The D_{nT} values are found in good agreement with that of the ISO standards. The detailed procedure is explained in Section **5.3.1**.

Now we create example cases with non-standard settings (i.e. configurations) of complex outdoor situations for scene 1 and scene 2 in order to evaluate the effects of source position and the complexity of the sound propagation model in the streets at different distance from the façades.

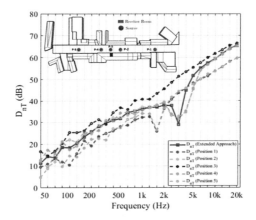

Figure 5.25: *Scene 1: Comparisons of standardized level differences D_{nT} between outdoor source and receiving room (all transmission paths and direct sound field only) of non-standard configuration settings for the source*

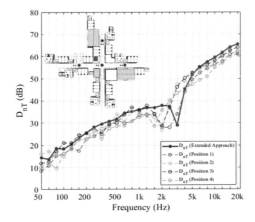

Figure 5.26: *Scene 2: Comparisons of standardized level differences D_{nT} between outdoor source and receiving room (all transmission paths and direct sound field only) of non-standard configuration settings for the sources*

This study is done to understand how big the changes in the effective D_{nT} values of the receiving rooms are in case of moving outdoor sources when the outdoor sound field includes only direct sound field and early reflections from the other dwelling in these scenes. Figure **5.25** and Figure **5.26** compare the results of both scenes such that in Figure **5.25** the source is placed at five different positions on the straight road facing at different angles towards the frontal façade while in Figure **5.26** the sound source is placed at the same number of position, however, located in different streets relative to the receiving room. In this case, the direct sound field is taken into account (i.e. without considering the reflections from the surrounding buildings). The differences in the D_{nT} values for the selected source positions might be because of the variation of the standard situation concerning distance and angle of incidence of direct sound field to different secondary sources (i.e. windows). It can also be observed that the dip in the curves for different positions of the source is shifting at different frequencies. Hence, the actual sound insulation curves differ from the standard results (ISO).

Now the complexity of the outdoor sound field is increased in terms of introducing early reflections of different orders to observe the changes in the level differences for different source positions. The direction of each arriving early reflection at the windows is taken into account and corresponding angle dependent transmission coefficients are applied.

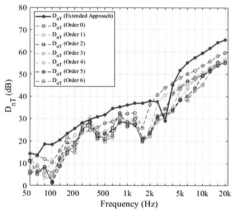

Figure 5.27(a): *Scene 1: Comparisons of standardized level differences D_{nT} between outdoor source **position 1** and receiving room (direct sound and reflections)*

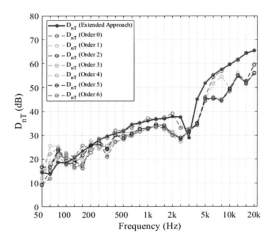

Figure 5.27(b): *Scene 1: Comparisons of standardized level differences D_{nT} between outdoor source **position 2** and receiving rooms (direct sound and reflections)*

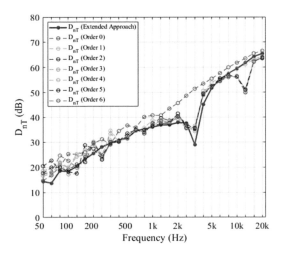

Figure 5.27(c): *Scene 1: Comparisons of standardized level differences D_{nT} between outdoor source **position 3** and receiving rooms (direct sound and reflections)*

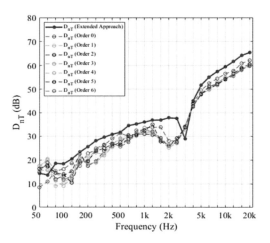

Figure 5.28(a): *Scene 2: Comparisons of standardized level differences D_{nT} between outdoor source **position 1** and receiving rooms (direct sound and reflections)*

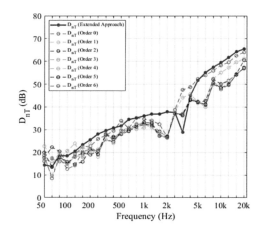

Figure 5.28(b): *Scene 2: Comparisons of standardized level differences D_{nT} between outdoor source **position 2** and receiving rooms (direct sound and reflections)*

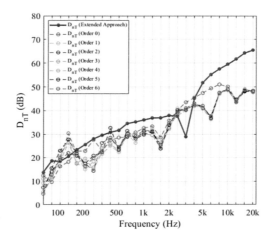

Figure 5.28(c): *Scene 2: Comparisons of standardized level differences D_{nT} between outdoor source **position 4** and receiving rooms (direct sound and reflections)*

Figure **5.27(a-c)** show comparison of D_{nT} for urban scene 1 for three source positions. For each position the sound field is calculated at façade elements with different orders of early reflections (order 1 to order 6). Similarly, Figure **5.28(a-c)** compares the results of scene 2 for three different positions. The differences in the D_{nT} values for one selected source position for different reflection orders do not show large variations. It can be observed that the standardized level does not vary with the complexity of the sound field, however, depends on the positions of the sound source and complexity of the façade.

In this chapter we evaluated the extended model of sound insulation filters by taking into account the source as well as receiving room acoustics with more details. The results of spatial variation of sound pressure inside the receiver room are presented using the knowledge of sound propagation theory in closed spaces for indoor and outdoor cases. Therefore, this model enables to experience more realistic loudness, colouration and binaural impression of the sound transmission at the receiving end by the sound source directivities and source and receiver positions in real-time. In addition, considering building elements as secondary sources might be helpful to include a more realistic directional cue of sound sources. In any case, measured sound transmission coefficients from test facilities may serve as input as well, so that existing (real) building situations can be simulated and compared with measurements.

Under conditions which match the measurement standards of sound insulation testing, the results of the auralization could be validated to differ not more than on average 0.68 dB and 0.3 dB for outdoor and indoor source positions, respectively. It is shown that in the results of the extended model, the source directivity and position have an influence on the transmitted energy to the receiving room and, thus, in turn the spatial variation of sound pressure level is more specifically related to the actual scenario and more valid when it comes to auralization. This fact is more obvious in the case of outdoor sound propagating to the receiver room through façades, where we can see that the secondary sources which are more exposed to incident sound field transmit more energy to the receiver room. In the next chapter the whole model is transferred into a signal processing domain which allows for interactive real-time processing and Virtual Reality applications.

6

Auditory-Visual Virtual Reality Framework

Audio stimuli are crucial for creating a convincing acoustic virtual reality (AVR) experience. When it comes to important acoustic features in perceptual studies, audio cues play a key role in our sense of being present in an actual physical space, so it contributes to the user's sense of immersion. In an AVR experience, we step into and are entirely immersed in a virtual world. This chapter discusses how to create and model AVR experiences concerning the building acoustical environments. Auralization of sound insulation of indoor and outdoor spaces is elaborated for real-time processing. We also discuss the basics of 3D graphics rendering tools that are vital for building acoustic implementation to create virtual built-up environments. Related audio rendering techniques are based on architectural design, geometry manipulation, real-time convolution, digital signal processing and sound insulation filters rendering. All this makes AVR environments to appear realistic and immersive. Above all, the main focus in this chapter is to pay attention to the particular requirements of AVR environments including performance issues and making sure that the sound insulation filters run fast enough in such environments. We make use of the professional game engine and VR software such as Unity software [63] and developed in it virtual building acoustics framework (VBA) as an open-source software package [61]. Unity is one of the most demanding game engines and is a relatively easy, however, fully featured.

The recent up-to-date developments in acoustic virtual reality [52,53,54,70] which are implemented in AVR environments [e.g. 55] are only available for room-acoustical simulations and auralization, especially developed for the closed spaces. Sound insulation auralization is not yet implemented in interactive audio-visual environments, e.g. the integration of sound insulation rendering and auralization for virtual built environments in AVR. To achieve this, three levels of implementation are associated with a universal platform for such kind of advanced virtual building acoustic frameworks. At first, the sound insulation prediction models are required to predict insulation metrics and are discussed in Chapter **4**. The sound insulation models and sound insulation rendering for auralization, in signal and filters domain, are the core parts of VBA framework. The sound insulation rendering and auralization involve digital signal processing techniques such as real-time convolution and binaural reproduction. The second level is the implementation of room and building acoustic filters as plugin concerning the real-time interactive audio-visual VR technology. The third level concerns visual rendering techniques for manipulation of the interaction of the user with VR environments and be present in the VR scenes which are being auralized. The main manipulation of the situation is the user's free movement in the scene. In this chapter, we discuss the implementation of VBA step by step in audio-visual virtual environments by integrating room acoustic and building acoustic filters, filter rendering methods, creating interactive scenes for different real world situations, such as office work sites and urban street scenes. We also discuss the auralization processing chain of VBA framework, its evaluation and real-time performance with example audio-visual scenes. Later on, in Chapter **7**, we will discuss the applications of VBA framework for evaluation of the performance of the buildings and designing listening experiments such as evaluation of background noise impacts on cognitive performance of humans under different building acoustical conditions and effects of intermittent outdoor moving sound sources on intension catching. Furthermore, we will discuss the example studies of listening experiment which allows the test subjects to perform any task of daily life of work or learning under conditions of usual behaviour and movement.

6.1. Virtual Building Acoustics (VBA) Framework

This section describes an audio-visual virtual reality framework (which is named as VBA) and its main components (called packages) for the virtual built environments, such as private dwellings, commercial office sites and urban environments. The development of VBA included Master's thesis projects [**82**] with a main focus on implementation of sound insulation filters for adjacent rooms. In [**82**] an initial idea for design of sound insulation filters was introduced based on the concept of subdivision into surface patches. In this dissertation, this concept was extended to advanced features for application on indoor and outdoor sound insulation auralization. The key features of VBA framework are given in Figure **6.1**. This framework enables us to perform room acoustic and building acoustic simulations in virtual reality environments and corresponding filters rendering and auralization of auditory-visual scenes. The pre-requisites to implement VBA into VR environments are;

1. Architecture model design tools
2. Appropriate virtual scenes rendering software and
3. Real-time audio rendering tools

These prerequisites provide an interaction with virtual environments and make realistic and immersive scenes which lead towards a plausible subjective evaluation of building performance and interaction.

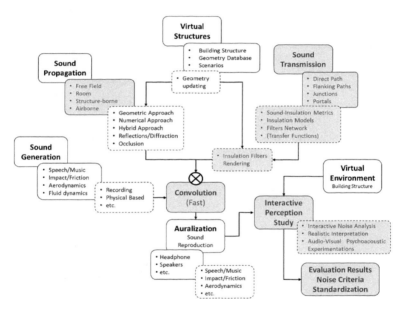

Figure 6.1: *Virtual building acoustics framework (VBA) features diagram*

As mentioned before, we used game engine Unity [63] as graphical (i.e. visual) rendering tool, whereas the geometrical models (i.e. architectural environments) are designed in Sketchup software [71] which provides geometric data as input for Unity. Unity is one of the most powerful game engines and visual objects rendering software which is relatively easy as compared to other available commercial software. It is fully featured with all necessary modules and packages that are pre-requisite the construction of virtual reality environments as a platform for the implementation of

VBA. It also supports the importing features of architecture geometrics directly from other third-party commercial software such as Sketchup [71] and AutoCAD etc. The building-acoustic simulation tools are made available for Unity as VBA-Packages which can be directly installed into Unity software. A complete documentation of VBA framework is available (www.virtualbuildingacoustics.org) [61].

6.1.1. Architectural Models

Architectural models are virtual built-up structures (i.e. the building with the geometric information) which are needed to be modelled with all its components and elements, for example, walls, doors, junctions and portals etc. For the evaluation and implementation of VBA framework, we created two virtual built up environments which are; 1) an office worksite and 2) urban street scenarios with two levels of complexity in architectural structures (discussed in Chapter 5). The first virtual building environment, which is an office work site, is the Institute of Technical Acoustics (ITA) RWTH Aachen. The geometry of the ITA building is taken as work-site premise for implementation of indoor building acoustics to auralize the sound transmission from an adjacent office room (from source room to an adjacent receiving room). The ITA building and its components (which are its walls elements, doors, junctions, floor and ceiling) are modelled in Sketchup software with actual physical dimensions but only with plausible input data, due to lack of precise information about the actual construction materials. Figure **6.2** shows different snapshots of ITA building designed in Sketchup software. Further snapshots are given in annex **A-9** for the interior designs of ITA building model. The second virtual built-up structures are two virtual urban environments consisting of different scenes which are designed from real urban street with different crossroad junction types. These scenes are **1)** street canyon and **2)** crossroad junction. These two urban scenes differ in complexities with regard to architectural designs of building constructions and the position of the source on the road relative to the facades to investigate the different aspects of outdoor sound field and its effects on sound transmission through façades. Figure **6.3** and Figure **6.4** show urban first and second urban scenes. Annex A-10 show different viewpoints of these urban scenes. Having detailed architectural models of both environments these models are exported to Unity® where rest of the virtual environmental components are designed and developed to best match with the real world environments.

Figure 6.2: *ITA-Building: External view of architectural construction model*

Figure 6.3: *Urban Environment **Scene 1**: External views of construction model (The green floor building is selected as case study)*

These virtual environmental components include graphical features to create and realistic visual part of the auditory-visual virtual reality framework.

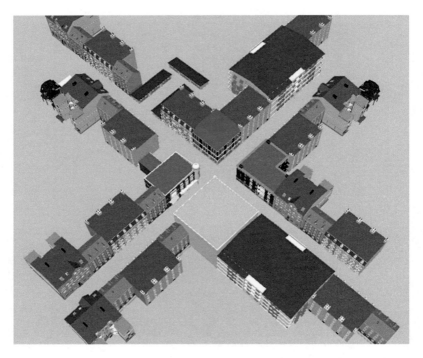

Figure 6.4: *Urban Environment **Scene 2**: External views of construction model (The green celling building is selected as case study)*

6.1.2. Virtual Reality Visual Environments

Once we have detailed model for architectural designs of the dwellings, these models are now imported into Unity® where rest of the virtual scenes are developed. The unity implementation of the environments (both ITA building and urban scenes) are shown in Figure **6.5** and Figure **6.6**.

Figure 6.5: *ITA building – two adjacent rooms used as example case for evaluation: External view (upper) and internal view (lower)*

Figure 6.6: *ITA – A classroom used as example case for perceptual evaluation of façade insulation (see Chap. 7): External view (upper) and internal view (lower)*

6.2. Implementation of VBA

The technical part of VBA framework is described in this section, which is composed of Unity packages developed in the form of Unity class functions and includes:

1. Room Acoustics Package
2. Building Acoustics Package
3. Outdoor Sound Propagation Package
4. Signal Processing and Audio Rendering Package
5. Geometry Manipulation and Visual Rendering Tools
6. Visualization Package

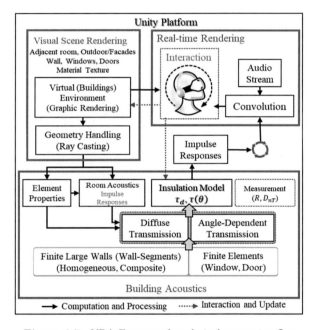

Figure 6.7: *VBA Framework technical processing flow*

The conceptual processing flow of VBA framework is shown in Figure **6.7**. In Figure **6.7**, there are three main computational parts of the VBA framework which are:

1. Building acoustics and room acoustics as audio rendering cues for the auralization

2. Visual rendering to simulate the corresponding virtual environments.

3. Real time auralization, Interaction and updating

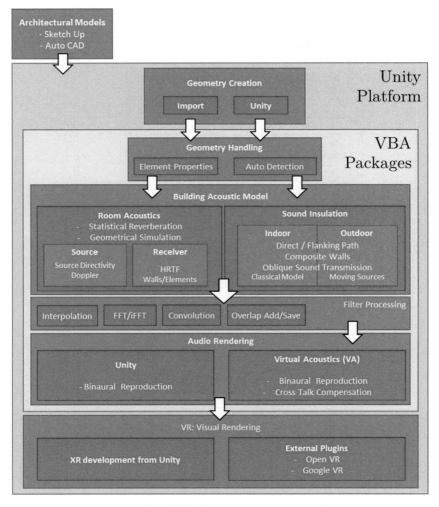

Figure 6.8: *VBA Packages and hierarchy chain*

The visual rendering includes built in Unity packages with that all the graphical features are integrated into virtual scenes, such as audio sources, the listener and the camera tools with interactive features to virtual controllers and the interactive scene (i.e. first-person controller (FPC)). The FPC tool is built-in Unity package. However, more up-to-date FPC tools with modern interactive features are also available online (e.g. Steam-VR) [75]. For the implementation of VBA interactive virtual environments, we used Steam-VR FPC tools. These tools help the user to interact with the VR scenes, visualize through HMDs and manoeuvre in VR environments, such as walking and moving around. Many audio rendering packages (i.e. Steam-VR, VA) are available for auralization which include general room acoustic effects such as room reverberation, audio mixers etc. However, these renderers support diffuse-field conditions only, which do not fulfil the requirements of simulating a complete building acoustics auralization for complex virtual architectural environments in general. One of the features of Unity is that is supports writing scripts in C# and JAVA languages for development source codes for customized functions to fulfil the requirements audio-visual rendering. After making use of this feature of scripting with C# language in Unity, all the room acoustics, building acoustics and audio rendering computation algorithm are developed as VBA package, installable directly into Unity platform. Figure **6.8** shows important functions (Unity class functions) and their interaction/relationship with overall computational hierarch which we will discuss and give an overview with key features in next section one by one. However, detailed documentations is given online at (www.virtualbuildingacoustics.org) [**61**], with complete open source VBA Packages.

6.2.1. Room Acoustics Package

At the first stage, to implement VBA framework for a given dwelling, either for indoor scene or for outdoor scenes, the room acoustics properties of source and/or the receiver rooms must be calculated to include its effects on sound insulation auralization. The room acoustical properties, which are normally described by room impulse response (RIR) alter the perception of sound transmission at the listener's end. The synthesis of room impulse response at surface elements of source room are necessary in order to include the effects of absorption and scattering from the room boundaries. It also is required to calculate the amount of incident energy on the wall surfaces which is then transmitted to the receiver room via direct as well as flanking paths. Therefore, room acoustics package deals with the room acoustics parameter and RIR synthesis that are important in auralization process and influence the room impression at listener's end. The main key features of this package include real-time simulation of room impulse responses synthesis at different positions using both statistical and geometrical approaches. In this work, simplified room acoustic model as described in section. **2.1.1** is used. But the geometric approaches may include image source method (ISM) and ray tracing (RT) techniques (the details of these method are given in [**54**]). The corresponding algorithm for simulations for ISM and RT are developed in C# language and included in VBA framework in the form of Unity-Class functions. Figure **6.9** shows example of implementation of room acoustics package for an office room situation using hybrid method (which is a combination of ISM and RT approaches) based on [**54**]. The details of simulation of this module are published at VBA website [**61**].

Figure 6.9: *Room acoustics module: ISM and RT methods (hybrid method) applied in a rectangular room for RIR synthesis in Unity platform. ISM: with order 3 is shown in yellow colour (audible only) and RT: with 1000 rays casted shown in blue colour (audible only)*

6.2.2. Building Acoustics Package

The building acoustics package calculates airborne sound insulation quantities based on sound insulation prediction models discussed in Chapter **4**. This package computes all the necessary input data for sound insulation filter design, given as follows.

1. Sound reduction index R (i.e. the one-third octave band frequency values)
2. Transmission coefficient in energetic form (i.e. the one-third octave band frequency values)
3. Normalized level difference D_{nT} of the individual building elements
4. Physical properties of the building elements
5. Radiation efficiency of the elements
6. Transmission loss factor
7. Vibration transmission loss at junctions

These quantities are calculated for both heavy weight and lightweight building constructions as given in ISO [6,17] and the sound insulation models described in Chapter 4. These insulation quantities are used as input data for filter construction. It is also possible to obtain input data for filter design process from product data sheets, from laboratory or from in-situ measurements for any specific building element (for example from databases in software like BASTIAN [72]). The building acoustics package includes different sound insulation prediction models which are discussed in detail in Chapter 4. These models are

1. ISO: 12354 (Part-I): (*Estimation of acoustic performance of buildings from the performance of elements: Airborne sound insulation between rooms*)
2. ISO: 12354 (Part-III): (*Estimation of acoustic performance of buildings from the performance of elements: Airborne sound insulation against outdoor sound*)
3. Classical simple model (Simple Adjacent Rooms)
4. Extended sound insulation model (adjacent room and outdoor sound sources)
5. Advanced sound insulation model {based on complex geometries of the building elements, urban environments}.

The advanced airborne sound insulation auralization model supports different methods and techniques (derived from [20,22,24,25,29]) for predicting sound insulation quantities for different building acoustical conditions, such as simple indoor situations (which includes adjacent rooms either separated through main partition or corners junction), complex indoor situations (rectangular source and/or receiving rooms with composite wall elements or partitions including doors etc.), simple outdoor situations (a single external wall as façade element) and complex outdoor situations (a corner receiving room with two external walls as façade elements). It handles simple homogenous as well as complex composite building elements. Figure 6.10 shows a snapshot of prediction of the sound reduction index (R') of the main partition for an adjacent room situation after implementation of the package. Few examples are given in Annex **A.10** for different indoor and outdoor (simple as well as complex) building acoustic conditions to illustrate the features of building acoustics package.

It is important to mention in this section that there are some differences in terms of accuracies concerning the building acoustics model part and the room acoustics model part of the framework. The room acoustics model used is more accurate than the building acoustics model as the building acoustics model is based on approximate energetic calculation without distributions of structural radiations which, however, is not the scope of this work to develop an accurate building acoustics prediction model rather to develop such a framework where the accurate building acoustics models may be integrated.

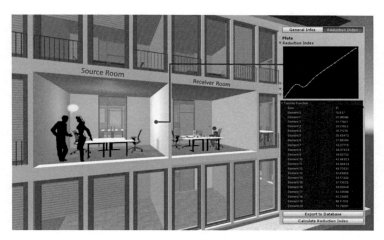

Figure 6.10: *An example of predicted sound reduction index (R′) for the partition (main separating element) between the adjacent rooms in VBA*

6.2.3. Outdoor Sound Propagation Package

The outdoor sound propagation package includes the sound propagation models for urban environments. It is applicable for auralization of virtual urban environments and for estimation of outdoor sound field (transmitted by an outdoor sound source) at façade elements of dwellings. This package contains two models which are,

1. **Virtual Acoustics (VA) Module:** VA calculates the impulse response from the source to the façade including the influence of the surrounding buildings using image source model (ISM) and image edge diffraction model.
2. **VAB Outdoor Module:** Computes the impulse responses using ISM and RT (ray tracing) method. However, it does not yet include the diffraction path contributions.

In both modules (VA/VBA-Outdoor), the reflected energies from the surrounding buildings are calculated for the façade elements from the impulse response and then transmitted to the receiver room through direct as well as flanking paths. Hence, the output from this module is an impulse response at the façade (at each façade element) which are further multiplied with the transmission coefficient of the façade elements of the receiver room. The examples of the simulation for outdoor sound propagation module are given in Annex **A.11**. Figure **6.11** shows an urban street canyon with implementation of outdoor sound propagation package as an example. The reflection and diffraction paths are calculated from VA software.

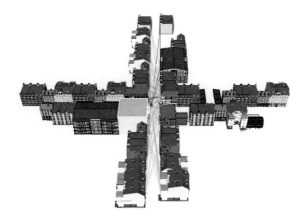

Figure 6.11: *An example of computation of sound propagation paths for a fixed source at the building façade*

6.2.4. Geometry Handling Package

The geometry handling package is the main component in the VBA framework, as all other modules, directly or indirectly, depend on it. This module is activated at the initialization step of the framework. It is based on the geometry of the built environment, and it computes all necessary input data for room acoustics and building acoustics simulations. This module automatically detects all building elements, such as walls, doors, junctions, portals etc. (including façades and their components in case of external wall elements of receiver rooms) after a virtual built structure is imported into the Unity platform from Sketchup or other CAD models. Once the built structures are imported, it is assigned source and receiver rooms (for adjacent room sound insulation). The physical dimensions (i.e. length, width, thickness, volumes, equivalent absorption areas of the rooms, reverberation time etc.) are automatically calculated using ray casting approach in the geometry handling module. The material properties such as internal transmission loss factors etc. are assigned to each element of source and the receiver rooms based on its physical and building acoustics characteristics.

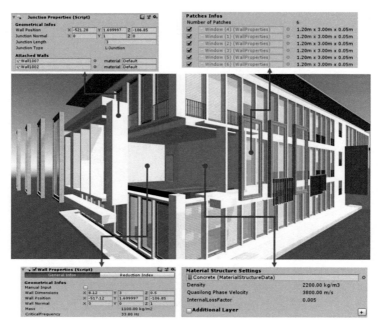

Figure 6.12: *Geometry handling: Assigning material parameters and interactive computation of geometric input data (in VBA)*

This package provides the basic input data for room acoustics package as well as for the building acoustics packages to compute room-acoustical characteristics and sound insulation filters respectively for the auralization. The material properties of the building elements can be predefined or assigned interactively to the elements via graphical user interface in VBA Framework as show in Figure **6.12**.

6.2.5. Transfer Function/Audio Rendering Package

Transfer functions are sets of filters from source to the listener, include;

1. Sound transmission filters (sound reduction index) for all transmission paths (for example direct and flanking paths)
2. Source directivity
3. Room Impulse Responses (RIR) filters of both the source and the receiver rooms (reverberation tail in case of diffuse field assumptions)
4. Head-Related Transfer Functions (HRTF) or corresponding time domain Head-Related Impulse Responses (HRIR).

These filters are used as input for audio rendering and auralization. At first, the transfer function package detects positions and orientations of the sound source and listener within the dwellings (i.e. source and receiving rooms, e.g. in case of adjacent rooms) and outside the dwellings (i.e. in case of urban scenes). This process is completed by using ray casting algorithm (Unity package). From an input time signal $s(t)$, which is obtained from the databases or recordings, and transfer functions (i.e. filters), the time signal at the output of any LTI system can be calculated by means of convolution techniques. This is done by open source FFT/iFFT libraries (FFTw) from MIT [**81**]. Hence, the time signal at receiver in the room can be calculated from the source signal and the transfer function from source to the listener.

As mentioned in Chapter **3**, the audible frequency range for human hearing is typically **20**Hz to **20**KHz, however, we generally consider the building acoustics frequency range to be defined by one-third-octave-bands from **50**Hz to **5**KHz or even only between **100**Hz and **3150**Hz. For signal processing, these quantities have to be turned into higher resolution of frequency spectra with a practical number of frequency lines in case of measured data. In case of classical sound insulation filters, an input with **21** values in one-third octave bands ranging from **50**Hz to **5**KHz is given. To obtain a frequency spectrum with **4097** lines, these values have to be interpolated. This is done applying cubic spline interpolation on insulation filters. In the receiving room, the building elements act as secondary sound sources located at different positions and orientation relative to the listener. Therefore, it is required to implement a perceptually correct localization in order to create a spatial impression of listening the room. Hence, it is necessary to consider binaural signals by using head-related impulse responses (HRIR) or corresponding head related transfer function, HRTF in the audio signal processing chain. The HRTF database measured from ITA dummy head is used in this thesis work. Thus, to simulate the spatial impression of the receiving room, simulated RIRs for the receiving room for each secondary source to the listener position are created. In this package, the simulated room impulse responses are convolved with the corresponding HRIRs. The main key features of this package are:

- Real-time source/receiver detection by using tracking devices
- Updating their positions and orientations in real time
- HRTF database updating from the relative angles of secondary sources with reference to the head orientation (view direction).
- Real-time FFT/iFFT based convolution

Further details can be found on VBA website [**61**], and audio rendering flow charts of this package are given in Annex **A.11**.

6.3. Evaluation of Real-time Performance (VBA)

This section evaluates the overall real-time performance of auralization of building acoustic filters for different cases, such as indoor and outdoor scenarios, for each step involved in terms of latencies. The main concept of real-time implementation is the construction of filters and their rendering in real time. The main algorithms included in VBA framework are geometry handling, sound insulation parameters calculation, computing transfer functions for secondary sources, including directional information using the HRTF dataset, and updating the filters for source and listener movement in real time. The complete framework rendering processes are categorized into two main part as

1. Offline filter calculations or initialization of framework
2. Real-time filter rendering and convolution process including updating filters

For calculating the latencies and performance of each process few example cases are taken into account, such as, indoor simple adjacent room case, outdoor façade sound insulation rendering case. All computations are performed on desktop computer featuring an Intel Core i7-7700 CPU @ 3.60 GHz multi-core with 16 GB RAM, Windows 10 (64-bit) operating system. The following scenarios are taken as example for calculation of latencies.

Figure 6.13: *Two adjacent rooms with partition and flanking walls, divided into patches*

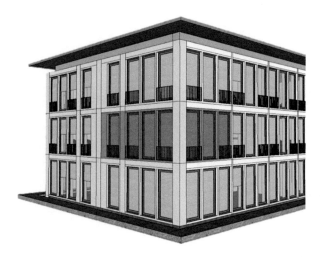

Figure 6.14: *Geometry handling: Assigning material parameters and interactive computation of geometric input data*

In Figure **6.13** and indoor adjacent rooms are selected with partition consists of two doors in a concrete wall. The doors and a portion of concrete wall serves as patches, total three patches on the partition. Similarly, the flanking wall is divided into three patches. In Figure **6.14** a corner room is selected consisting of two façades with windows as patches. There are total seven patches.

6.3.1. Filter Construction (Initialization)

At first, the geometry handling process takes place which use the Geometry Handling Package. This process makes the geometric data available for calculation of sound insulation metrics (by using Building Acoustics Package). This process is performed offline, and hence is required only for initialization of the VBA. The computational cost for this process is **15.2**ms for indoor scenario whereas it is 20 ms for outdoor scenario. Secondly, sound insulation metrics for each element of the building are calculated and subsequently transfer functions for each path (direct as well as flanking) are computed in terms of transmission coefficients spectra in one-third octave band with a frequency range of **50**Hz to **5KHz**. These spectra are interpolated by using a **4097** points cubic spline interpolation to get suitable audio filter and to extend the range of an appropriate frequency resolution for the selected coupled rooms. Afterwards, these spectra are used to calculate the final transfer functions from the sound source to the secondary sources by adding the different flanking transmission paths. The calculation time required for seconds step is **20**ms, for three patches, however, this process works interactively by just clicking on building elements in Unity [**63**].

6.3.2. Real-time Filter Rendering and Convolution

Once the building acoustics filters and HRTFs filters are available as input data, real-time filter rendering and convolution process for these filters is performed using the Audio Rendering Package. The first step is to synthesize impulse responses from the source to the façade elements (in case of urban environments) and from the source to the partition and flanking elements (in case of adjacent rooms). Afterwards, these IRs are convolved with sound insulation filters. This way, the virtual sound is transmitted to receiver room. Inside the receiving room, each wall and/or façade element is considered as independent secondary sound source which radiates the sound to the listener in the room. The position of these secondary sources relative to the listener are calculated and HRTFs are applied between the direct sound field component of the secondary sources and the listener by simply multiplication of the frequency spectra of each. The receiver room impulse response is then convolved to include the room response. An inverse Fourier transform (IFFT) and summation is processed to get final binaural signal in time domain. These signals represent multiple independent paths (depending on the number of secondary sources). Therefore, they are amplitude-weighted and delayed according to the dimensions of the receiving room and the position of the listener with respect to them. As a last step, the final binaural signal is convolved with an example sound signal for auralization. The sound file has typically 24-bit digital resolution and 44.1KHz sampling rate. This operation is performed by FFT/IFFT (including overlap add method) by dividing the source signal into frames of 256 samples, transforming to frequency domain, multiplying with binaural filters and converting back to time domain. Figure **6.15** shows the computational cost for real-time processes. With this, interactive scenarios can be created in which the listener can freely move (turn the head or walk around) in the virtual receiving room.

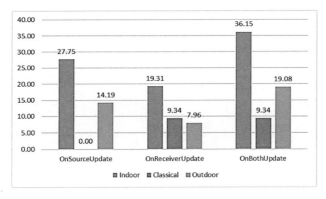

Figure 6.15: *Computational cost of each model for each real-time process (in ms)*

6.4. Audio-Visual Scenes

The usage of VBA framework in audio-visual environments is as a platform for psychoacoustic and cognitive performance experiments under desired building acoustic conditions. With the VBA framework, novel listening experiments on evaluation of noise effects on humans in built-up environments are possible to be performed which go beyond the traditional listening experiments. Typical listening experiments require reproducible conditions, which can only be guaranteed with simplifications in terms of the stimuli and the context of the test in a so-called "laboratory" setting. With listening experiments designed in VBA, the reproducibility is maintained, while the features of the acoustic stimulus and the situational context can be enhanced by:

- More realism due to audio-visual scene rendering
- Immersion of the test subject in the virtual scene

- Almost natural behaviour of the test subject in the environment
- Interaction (free movement of sources and receiver)

Figure 6.16: *Audio-Visual scene for listening experiment*

This opens large opportunities for progress in hearing research and noise effects research. A listening experiment requires a test paradigm which explains the tasks for the participants. Examples for designs of listening experiment paradigms are explained in Chapter **7** for real-time sound insulation auralization and for the perceptual evaluation of noise stimuli with more realism and more contextual features. Here we show a few snapshots of these listening experiment paradigms and the important features of the software and its usage to present different background building acoustical conditions to participants. Several room acoustics and building acoustics features are included in VBA framework which can be switched on/off during a particulate listening experiment. For example, activation and deactivation of room acoustics features (such as reverberation) and the selection of room acoustics methods, source and receiver characteristics (such as position, orientation and directivity) and most important the choice of the methods for sound insulation filter design and rendering technique. These options can be selected during the configuration and initialization of the VR environment. The VBA framework details and documentation are published online at (www.virtualbuildingacoustics.org) [61] with examples and demos to explain its usage for audio-visual virtual reality environments, however, Figures **6.16** shows a snapshot of the audio visual virtual environment.

7

Perceptual Studies

The performance of buildings concerning protection against noise can be evaluated from a technical perspective as well as in a human-centred approach by considering subjective ratings, cognitive performances, or other human activities. The technical-oriented evaluation is based on the standard measurement and prediction techniques. Laboratory and field measurements, as well as prediction models for planning purposes, are quite advanced. The evaluation and single-number rating, i.e., the so-called "sound insulation metrics" are used in practice to ensure a proper noise protection, for which limits (requirements) are set by national authorities [**6,12,16, 85,86**]. The measurement, prediction and decision about noise control in building acoustics are based on the correlation of the sound insulation metrics with subjective ratings and with field surveys at the national level [**4**] with different metrics and different noise limits. Studies involving subjective ratings are performed mostly with direct assessment of the noise effect (loudness, annoyance) in psychoacoustic experiments. These tests focus on the auditory stimulus that draws the test subject's attention and concentration to the sound event [**56,57**]. Here, the question that often arises inquiries about the extent to which the implausible laboratory situation reduces the validity of the obtained results. Moreover, questionnaires have been used in the field to collect the data on noise effects from inhabitants of a building or a city quarter. The problems with survey data include memory effects and ambiguities of making a statement about annoyance concerning longer periods (such as days or weeks). In contrast to these sound-focused procedures applied in the laboratory or in the field, in real life, background noise from neighbours, building equipment, and traffic is present while people are engaged in activities such as working, learning, resting, and others. Here, noise and its evaluation are not in the foreground of the real-life human activity. Behavioural tests assessing cognitive performance tasks to be worked on during background noise, may give a more reliable insight into noise effects on humans [57]. Cognitive performance loss, for instance, can be interpreted as an index of performance loss in working or learning environments. Although the effect of office noise and in particular background speech *within* an office is the subject of research of several work groups [58,59].

In this chapter, at first hand, we design few psychoacoustics experiments to validate VBA framework and to illustration its potential applications for such listening experiments with interactive auditory-visual VR environments and on the other hand to evaluate speech and background noise effects on human cognitive performance under a variety of building acoustical conditions. We made use of the VBA framework as discussed in previous chapters in detail for such listening experiments in virtual reality environments. The first experiment is about cognitive performance of humans during background noise under good and bad speech intelligibility conditions of adjacent rooms. In the second experiment, we evaluate the perceptual localization capabilities of human due the outdoor moving sound sources. The second experiment is rather simple with respect to the outdoor sound propagation model and building construction complexities. In next sections of this chapter we chronologically discuss these listening experiments and its outcomes. Hence, plausible auditory-visual scenes allow to introduce more realism and more contextual features into psychoacoustic experiments These exemplary studies presented promise new options for research on noise effects by the use of virtual built environments which are of high plausibility and unlimited variability.

7.1. Cognitive Performance during Background Noise Effects[1]

In a laboratory listening experiment described in [5], the authors tested how different background speech conditions from an adjacent office affect cognitive performance in terms of short-term memory. This approach, however, could still be improved by adding plausible video input, considering the variability of the auditory-visual setting, reducing behavioural restrictions placed on the participants (e.g. no natural body movements, more or less static head position etc.), and varying the types of the rooms (e.g. office, workroom, living room and classroom etc.). This is the starting point of this section, where we introduce the first application of VBA framework in virtual reality for a psychoacoustic experiment. This psychoacoustics experiment was conducted during master thesis project [84]. The purpose is to investigate the effect patterns of the different background sounds under different building acoustics conditions for indoor scenes and compare results with that of real-life laboratory listening experiments[1]. With this, it is to validate VBA as an audio-visual virtual reality framework as a potential tool for exploring sound effects of building characteristics for different real life activities.

7.1.1. Building Acoustics Model (Adjacent Office)

The first prerequisite for this listening experiment in virtual reality environments is the building acoustical model of adjacent office rooms. In [5], the speech stimuli were created by a static sound insulation auralization and presented via headphones. They selected four sound conditions, a normal speech signal (speech level **55dB** L_{Aeq}, perfect intelligibility), two auralized versions of speech signals characterised by good intelligibility (speech level **35dB** L_{Aeq}, high frequency contents were large) and bad intelligibility (speech level **35dB** L_{Aeq}, high frequency contents were normal and attenuated) and a silence condition represented by very soft pink noise (**25dB** L_{Aeq}). The previous audio-only experiments from [5] showed that highly intelligible background speech impairs cognitive performance irrespectively of level (**55dB** vs **35dB** L_{Aeq}) and that a reduction in speech intelligibility is prerequisite for reducing adverse effects even of the soft speech signal. The values of weighted apparent sound reduction index R'_w for the "good" and "bad" insulating walls of [5] were approximately **37dB** and **26 dB**, respectively.

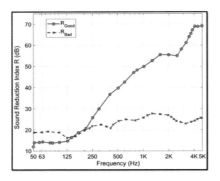

Figure 7.1: *Sound insulation curve for "good" (blue: heavy concrete wall) and "bad" (red: light timber) sound insulation for the adjacent offices corresponding to "bad" and "good" intelligible speech conditions, respectively, from [5]*

---

[1] This section 7.1. was published in [8]

For the listening experiment described in this chapter, we used an extended sound insulation model from Chapter 4 for which we select the source and receiver room walls in the adjacent offices in such a way that the final sound insulation curves for both good and bad walls are exactly the same as in Figure 1 of [5], in order to test the same sound conditions in this experiment in VR environment. Figure **7.1** shows these final sound insulation curves between the adjacent offices. In this way, all of the acoustic conditions were maintained exactly the same to facilitate the comparison between the audio-only experiment of [5] conducted in a real laboratory and the VR-based experiment with rendering of an audio-video office environment in the present study. The VR scene for the experiment is discussed in next section.

7.1.2. Virtual Reality Environment (VR-Scene)

As discussed before, the purpose of this study is to investigate whether the effect patterns of the different background sounds obtained in the laboratory listening experiment of [5] are reproduced in a VR environment presented by using the VBA framework through a Head Mounted Display (HMD) and by using binaural headphone technology.

Figure 7.2: *Source and receiving room: Office situation in receiving room with work desks and computer screen and persons talking in source room*

The real environment listening experiment in [5], where authors presented task stimuli on a real computer screen in a real room, is termed as "real-scene" in this section. Now, the same "real-scene" listening experiment is presented in VBA

Framework on a virtual computer screen in a virtual office room. This virtual office room is selected in the ITA building which is designed in VBA with plausible building acoustics features (see Chapter 6 for detail description of VBA Framework). This framework allows the user to interact with the environment such as free movement in virtual office room and interaction with virtual listening experiment equipment's (e.g. computer). The only *real* (i.e. non-VR) device in this present study is a computer mouse used by participants to complete the administered performance tests and provide subjective ratings. Figure **7.2** show the snapshots of the VR scene of adjacent offices. Reproducing the results of the previous study [**5**] would open a very large opportunity for using cognitive performance tests and evaluations by subjective ratings in VR with variable sound insulation settings, other background noises and in the context of different visual and more plausible daily-life (virtual) environments.

Figure 7.3: *Experimental setup in VR lab: Participants wearing HMD, visualizing the virtual receiving room office. HMD view point of the office room is shown on the projector screen (same as viewed in HMD)*

The specific question in this experiment was if frequency-specific sound insulation and speech signals transmitted through the wall creates impact on the cognitive performance, in particular in comparison of intelligible and non-intelligible speech. In the scene, the participants were presented a virtual reality environment, where they are sitting in an office-like receiving room, performing a cognitive task of verbal serial recall on the screen and hence evaluating the verbal short-term memory capacity. Their answers were recorded under different background stimuli of "irrelevant speech", originating from the neighbouring office (source room). The test is designed to study the impact of meaning of background sound (intelligible speech vs. non-intelligible speech) on the cognitive performance, the sound insulating components of the building are selected in the way that the final binaural signal at the listener's ear presents bad and good sound insulation. To present the scene to the participants for this in VR, the head-mounted display type "HTC Vive" [62] is used in Virtual Reality Lab (ITA) as shown in the Figure **7.3**. The headset uses "room scale" tracking technology, allowing the user to move in 3D space and use motion-tracked handheld controllers to interact with the environment.

7.1.3. Evaluation of VR environment: Cognitive performance and subjective ratings

The present experiment used the described audio-visual VR environment to explore the effect of background speech of differing intelligibility and level on verbal serial recall which is a standard procedure to measure verbal short–term memory capacity. The audio-visual VR experiment aimed to replicate a laboratory experiment conducted by [5]. In the previous study, the participants worked on the serial recall task on a real notebook's screen and heard the different background sound conditions via headphones without any corresponding visual input. Figure **7.4** shows a screenshot of the present experimental VR setting visualized through HMD view.

7.1.3.1. Methods

Twenty students, including 14 males and 6 females, of Institute of Technical Acoustics (ITA), RWTH Aachen University, participated in this experiment. They were aged from 21 to 37 years ($M_d = 26$ years), and they were all native German speakers. Audiometry tests of the participants were conducted to ensure that they had normal hearing. The experiment was carried out on a personal computer with an Intel Core i7 configuration (16 GB RAM). The visual office scene, developed in Unity [63] with building acoustics audio plugin [61] was presented to the participants through HTC Vive headset [62] as shown in Figure **7.2**. In the audio-visual scene, a typical office workplace was visualized consisting of office furniture, a mouse, and a computer screen, as shown in Figure **7.3**. In this virtual screen, the same verbal serial recall task with the same stimulus and timing settings was used in the previous laboratory listening experiment [5]. Digits from 1 to 9 were presented visually in the middle of the virtual computer screen (**700 *ms*** on, **300 *ms*** off time), replicating the previous procedure.

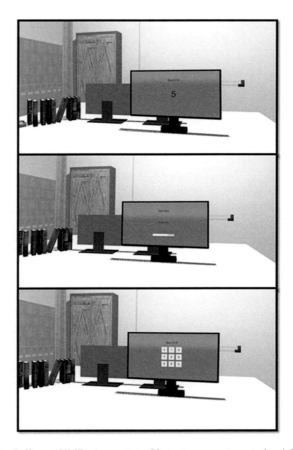

Figure 7.4: *Different HMD view point of listening experiment: (top) Random nine digits are displayed, (centre) waiting time to answer (bottom) answer panel*

The background sound conditions were auralized and played-back in such a way that highest similarity to [5] was provided. Speech conditions were derived from the same anechoic recordings of 130 short German sentences (5-6 words). These were semantically meaningful and spoken by a male speaker. A pause of **500** to **700** ms was given between sentences (the standard gap between sentences in common narration). In the present study, like previously, this spoken material was auralized in the receiving office at a level of $L_{eq} = 55$ dB(A) and was thus highly intelligible ($S55$). Two soft speech signals ($L_{eq} = 35$ dB(A) resulted from auralizing the spoken material in the source office and transmitted to the listener in the receiving office. Two differently shaped sound insulation curves representing unequal building situations [see Introduction, Figure **2** and [**5**]], one soft speech signal of good intelligibility (*S35_G*) and the other one of bad intelligibility (*S35_B*). Finally, a silence condition (*PN_25*) represented by very soft pink noise ((L_{eq} = 25 dB(A)) was included in the experiment to measure baseline performance of participants in the short-term memory task. In the present VR-based experiment, all background sounds were presented binaurally using Sennheiser HD 650 headphones. The sound pressure levels refer to an energy-equivalent sound pressure level L_{eq} averaged over presentation duration and measured using an arti cial ear (HMS-III, dummy head from HEAD Acoustics) and a sound level meter (Norsonic Sound Analyser 110) and calibrated the SPL for each sound condition through RME Fireface UC II sound card to deliver the target SPL to the headphones. Subjective ratings of disturbance, annoyance, task difficulty, ability to concentrate and effort were measured on a five-point scale with (**1**) = 'not at all', (**2**) = 'a little', (**3**) = 'middle', (**4**) = 'rather' and (**5**) = 'extremely' (in German: 'gar nicht', 'kaum', 'mittelmäßig', 'ziemlich', 'außerordentlich').

The procedure of the present VR-based experiment was kept as similar as possible to that applied in the real laboratory environment by [**5**]. The present experiment was conducted in individual sessions in a soundproof room inside the Hearing Booth 1 (U107) at the Institute of Technical Acoustics (ITA), RWTH Aachen University. It lasted approximately 60 min for one session. In written instructions, the participants were asked to ignore the background sounds and to work as fast and as accurate as possible on the short-term memory task. At the beginning of the experiment, five practice trials were given under silence condition (*PN_25*). In each trial, three rectangles decreasing in size (**1** *sec* onset-to-onset interval) indicated the upcoming start of digit series presentation on the virtual computer screen. Subsequently, the digits from 1 to 9 were shown in random order. After a pause of 10 seconds (retention interval), nine squared pushbuttons in the form of a 3 × 3 matrix appeared on the virtual screen in which the nine digits were randomly arranged. The participants were expected to reproduce these digits in exact presentation order by clicking on these pushbuttons using a virtual mouse cursor, which was moved by the use of a real wireless mouse. After clicking on a digit, the digit and the corresponding pushbutton disappeared and could not be selected again. Therefore, it was not possible to skip a serial position or correct errors. During each background sound condition, 12 successive trials had to be completed in one experimental block. A succession of sound conditions was balanced over participants. A pause of 2 min was given between sound conditions. Subjective ratings were collected after each background sound condition, as done in [**5**].

7.1.3.2.   Results

The behavioural data collected in the present VR experiment were analysed and compared with the data collected in [5]. The goal was to compare the noise effect patterns obtained in audio-video rendered office environment (VR) with those found in the real laboratory testing environment.

7.1.3.2.1.   Performance Measurements

In the first step, we tested whether the effect of background sound conditions on short-term memory differed depending on whether participants worked on the serial recall task in the VR environment or in the real laboratory setting. Since the two experimental groups differed in their performance baseline with respect to the serial recall performance during silence (VR: $M = 22.9\%$, $SE = 2.9\%$; Real: $M = 29.2\%$, $SE = 2.1\%$), difference values were considered in the following data analyses. It needs to be mentioned that such baseline differences between experimental groups are not unusual or "quite normal" if sample sizes are small, as in the present experiments ($n1 = n2 = 20$). Difference values were calculated for each participant by computing the individual difference in error rates during each background speech condition relative to the individual baseline performance during silence. Individual difference values were then averaged over all participants within each experimental group (VR vs. Real) to obtain group means relative to the respective performance during silence (Figure **7.5**).

To determine a significant difference in the sound effect pattern between the two experimental groups, a 2×3 Analysis of Variance (ANOVA) was conducted with the *environments* (VR, Real) as the between-subjects factor and *sound* ($S55$, $S35\_G$, $S35\_B$) as the within-subject factor. Most importantly, the interaction effect of *environment* and *sound* was not significant, $F(2,76) = 0.43, MSE = 0.002, p = .65, partial\ \eta^2 = 0.01$ as was the main effect of *environment* on error rates $F(1,38) = 0.22$, $MSE = 0.006$, $p = .64$, $partial\ \eta^2 = 0.01$. Furthermore, the effect of the background speech conditions on cognitive performance did not vary between VR and real laboratory environment. The main effect on *sound* is highly significant $F(2,76) = 17.10, MSE = 0.100, p < .001, partial\ \eta^2 = 0.31$, and is analysed in detail in the following. Although the statistical results from the joint analysis of VR and laboratory performance data are clear-cut, we conducted paired t-tests on the performance data collected in the audio-video VR environment to test in detail the observed sound effect pattern. Here, just as in the real laboratory environment [5], error rates during highly intelligible background speech (irrespectively of level, i.e.

$S55$ and $S35\_G$) were significantly enhanced compared to background speech of bad intelligibility $(S35\_B)$ and silence, $t(19) \geq 2.85, p \leq .01$, Cohen's $d \geq 0.45$. The latter two sound conditions did not differ, $t(19) = 0.50, p = 0.62$, just like the two background speech signals of good intelligibility, $t(19) = 1.85, p = .08$.

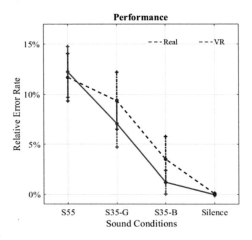

Figure 7.5: *Verbal short-term memory capacity in VR and in a real laboratory during highly intelligible speech at 55 dB(A), (S55) and auralized speech at 35 dB(A) of either good (S35_G) or bad intelligibility (S35_B). Mean error rates relative to silence difference values*

7.1.3.2.2. Subjective Ratings

The subjective ratings were analysed statistically analogously to performance data. By doing so, the same effect pattern was observed for all 5 subjective rating scales (Figure **7.6(a-e)**). Most importantly, based on subjective ratings, the interaction between *environment* and *sound* was non-significant, indicating that subjective ratings on the corresponding scale did not vary between VR and real laboratory environment $F(3,114) \leq 2.50, p \geq .08, partial\ \eta^2 \leq 0.06$. Furthermore, there was no main effect on the *environment* for any of the subjective ratings $F(1,38) \leq 3.30, p \geq .08, partial\ \eta^2 \leq 0.08$, but a significant main effect of *sound* on each of the subjective rating scales was found, $F(3,114) \geq 12.87, p < .001, partial\ \eta^2 \geq 0.25$. As panels B to F of Figure 8.3 depict, the same sound effect pattern emerged in both testing environments (VR and real laboratory) for all rating scales with $S55$ reaching the highest rating values, always followed by $S35\_G$, $S35\_B$ and finally silence with the lowest ratings.

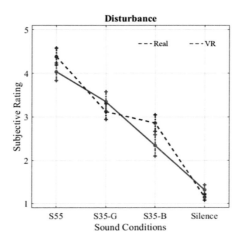

Figure 7.6(a): *Subjective ratings (**Disturbance**) in VR and in a real laboratory during same sound condition of performance ratings. Rating scales ranged from 'not at all' (1) to 'extremely' (5)*

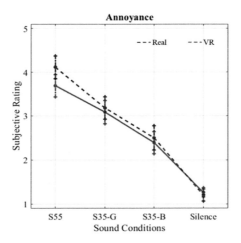

Figure 7.6(b): *Subjective ratings (**Annoyance**) in VR and in a real laboratory during same sound condition of performance ratings. Rating scales ranged from 'not at all' (1) to 'extremely' (5)*

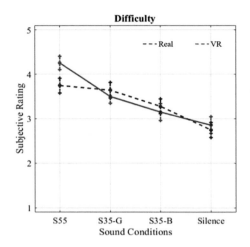

Figure 7.6(c): *Subjective ratings (**Difficulty**) in VR and in a real laboratory during same sound condition of performance ratings. Rating scales ranged from 'not at all' (1) to 'extremely' (5)*

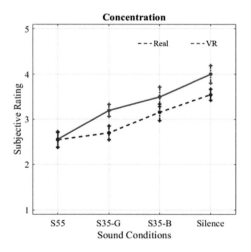

Figure 7.6(d): *Subjective ratings (**Concentration**) in VR and in a real laboratory during same sound condition of performance ratings. Rating scales ranged from 'not at all' (1) to 'extremely' (5)*

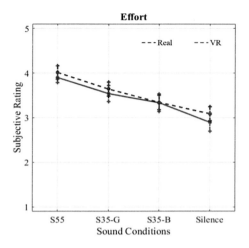

Figure 7.6(e): *Subjective ratings (**Effort**) in VR and in a real laboratory during same sound condition of performance ratings. Rating scales ranged from 'not at all' (1) to 'extremely' (5).*

7.1.4. Summary

The potential of using a human-centred approach integrating audio-video VR to evaluate indoor noise protection by building characteristics was exemplarily demonstrated in the present study. Different background speech conditions were derived from convolution with sound insulation filters of adjacent office rooms and presented in a VR office environment. The sound effect pattern on cognitive performances - in terms of verbal short-term memory - and subjective ratings were the same as those measured in a real and audio-only laboratory setting utilizing the same speech conditions [5]. With this, the study was the first to validate audio-video VR test environments as a potent tool for exploring sound effects of building characteristics on cognitive performances and subjective evaluations. The presented listening experiment with these configuration of the framework provides real-time auralization of sound insulation integrated with VR to evaluate the effects of building acoustic performance with respect to cognitive performance and subjective evaluations in a more plausible procedure. Although the presented auralization framework incorporates many important room and building acoustic effects based on our extended approach, sound presentation in the behavioural experiment was kept simple; i.e. static source and receiver in both source and receiver rooms.

In the next section, we used more dynamic auralization scene where we explored the perceptual localization capabilities of human due to the outdoor moving sound sources under building acoustical conditions. Furthermore, we aim to address outdoor moving sound sources to investigate the impact of intermittent dynamic noise of passing-by cars, ambulances or police sirens with subjective evaluations and other daily life activities (e.g., resting) while allowing the participants to engage in usual spatial behaviour and body movements.

7.2. Perception of Passing-by Outdoor Sources

As a next step, another listening experiment is designed to investigate perceptual localization capabilities of human under building acoustical conditions during background noise from outdoor moving sound sources. Specifically, the aim of such an experiment was to validate the real-time building acoustic rendering framework for façade sound insulation introduced in Chapter 4. This insulation model is used to construct insulation filters for sound transmission of moving outdoor sound sources by façade elements. The first innovative aspect is to implement a moving sound source in real-time auralization. Secondly, the objective of the listening experiment was to validate a hypothesis which states that in real built environments it is possible to perceptually localize the moving outdoor sound sources from inside the buildings.

7.2.1. Building Acoustical Model (Façade Sound Insulation)

Sound transmission from outdoor sound sources (e.g. a moving vehicles) into a building is a complex process as these sources, generally, are directional sources with strong low frequency sound characteristics. Mostly, the exterior walls of common buildings consist of an assembly of two or more parts or surfaces (e.g. windows etc.), therefore, the approach of segmenting the façades into finite size patches representing the window elements is applied. The description of outdoor sound insulation model and filters designing are described in Chapter 4. In this particular listening experiment, the only direct sound transmission paths Dd, through each secondary sound source are considered because it is assumed that the transmission for each sound source is independent from the transmission of the other sound sources. Therefore, for a direct sound field the sound transmission coef cient of a plane wave depends on the angle of incidence θ, between the direction of propagation of the incident plane wave and the normal to the plane of the exterior elements (i.e. façades). A corner classroom at ITA-building is taken as receiving room for the test case. The dimensions of the classroom are $8.12 \times 13.5 \times 3$ meters. The selected external walls (i.e. façade) of this classroom are an assembly of different components, i.e. one external wall consists of six single glaze windows, whereas, the other external wall consists of eight windows connected through concrete pillars. The height and width of each window are **2.5 m** and **1 m** respectively. Each glass thickness is **8 mm**, density is $\mathbf{2500 \frac{kg}{m^3}}$, and the internal loss factor is **0.001**. Each window act as secondary source for the receiving room and sound insulation filter for each secondary source are computed independently while considering them as finite segments. In this way, the façades of the classroom act as many secondary sources which radiate different sound energies to the receiver, placed somewhere in the classroom. The radiated of transmitted energy from each secondary sound source is depends on the amount of energy received by them from the source and their position and orientation relative to it.

Once the filters for sound insulation are calculated, auralization makes the sound pressure in the receiving room audible to listener by appropriate reproduction equipment. From an input time signal $s(t)$ and transfer functions (i.e. filters), the time signal at the output of any LTI system can be calculated by means of convolution techniques. Hence, the time signal at a receiver in the room is calculated from the source signal (in this case we selected a motorbike as noise source) and the transfer function from all secondary sources to the listener by convolution. In the receiving room, the secondary sources located at different positions and orientation relative to the listener's ears are excited by the sound transmission from the source, therefore, they are required to be perceptually localized to create a spatial impression of listening the room. Hence, it is necessary to consider an auralization with measured or individualized binaural signals by head related transfer function. Furthermore, to experience the impression of the receiving room, we have simulated impulse responses for the receiving room for each secondary source to the listener position. To make the listening experiment simple it is assumed that the reflections and the diffraction from the surrounding buildings and the ground are not influencing the sound power hitting on the surface of the façade. Therefore, only the direct part of the sound field is taken into account for auralization. The final sound insulation curve in one-third octave bands is given in Figure **7.7**.

The sound stimuli were created by a dynamic sound insulation auralization and presented via headphones. The selected sound source was an engine noise of a moving motorbike at the speed of 40 $\frac{km}{h}$ and produced a noise signal of **80dB** L_{Aeq} in a distance of 4 m. The listening experiment was conducted in both real and virtual environments. The virtual classroom architectural elements (i.e. walls and windows) are constructed in the way that the final normalized level difference D_{nT} of both real and virtual classrooms are exactly the same. In this way, all of the acoustical conditions were maintained exactly same to facilitate the comparison between the real environment experiment conducted in a real classroom and VR-based experiment with rendering of an audio-visual classroom. Reproducing the results of real environment would open another large opportunity for using cognitive performance tests and evaluations by subjective ratings in VR with variable sound insulation settings, other background noises and in the context of different visual and more plausible daily-life (virtual) environments. The VR scene of experiment is discussed in next section.

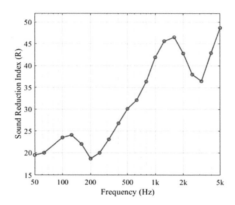

Figure 7.7: *The sound insulation curve for double glaze windows*

7.2.2.   Virtual Reality Environment (VR-Scene)

The purpose of this study is to investigate whether the outdoor moving sound sources can be perceptually localized inside the dwellings under sound insulation conditions obtained in real buildings rooms. Also, it is validated if this effect can be reproducible in a VR environment. In the real environment, we presented sound stimuli in a real classroom which is termed as "real-scene". The "real-scene" is then reproduced in virtual reality environment and same sound stimuli (real recordings) were presented to the participants, now in a virtual classroom. This virtual classroom is designed in the same fashion as the real classroom of ITA building which is designed in VBA framework with all its building acoustics features. Figure **7.8** shows the virtual classroom which is located at the ground floor.

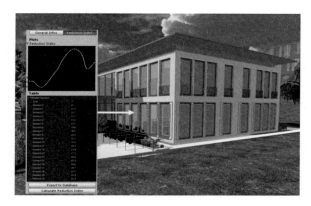

Figure 7.8: *A corner receiving room: Virtual classroom*

The specific question in this experiment was if frequency-specific sound insulation and outdoor moving sound signals transmitted through the façade creates impact on the localization of source, in particular in comparison of real and virtual environments. In both real as well as virtual scenes, the participants were presented same environments, where they are sitting in classroom as receiving room, performing a task of evaluating the direction of moving motorbike. The interior view of the virtual classroom is shown in Figure **7.9**.

Figure 7.9: *Virtual receiving classroom interior view. Four selected positions for presenting sound stimuli. The test subjects are facing the exterior wall to the street.*

151

The answers were recorded under two background stimuli of "motorbike", moving on the street on the left side in front of the façade from left to right and right to left randomly. To present the scene to the participants for this in VR, the head-mounted display HTC Vive [62] is used in Virtual Reality Lab (ITA). The headset uses "room scale" tracking technology, allowing the user to move in 3D space and use motion-tracked handheld controllers to interact with the environment.

7.2.3. Evaluation of VR environment: Perceptual Localization of Moving Outdoor Sources

The present experiment used the described VR environment to explore perceptual localization capability of human under façade sound insulation conditions. The audio-visual VR experiment intended to see that whether outdoor moving sound source can perceptually be localized due to frequency-specific façade sound insulation conditions. It aims at finding the correlation between the perceptual localization results of the outdoor moving source in real and virtual environments. The participants worked on a localization task in a real classroom at different positions, hearing the sound conditions via headphones as shown in Figure **7.7**. The same real environment experiment is simulated through VR setting and visualized through HMD. The audio-visual VR experiment aimed to compare the results obtained by conducting the same experiment in real classroom of same configurations.

7.2.3.1. Methods

A total of seven participants (5 males and 2 females) took part in the experiment. They were aged from **21** to **37** years ($M_d = 26$ years). Two auralized versions of sound conditions were included in both real and VR experiments, i.e. a motorbike running at $40\frac{km}{h}$ from left to right and from right to left in front of the façade with direct sound field component. Figure **7.10** shows the trajectories of the moving motor bike in front of the building façades.

Audiometry tests of the participants were conducted to ensure that they had normal hearing. The VR experiment was carried out on a personal computer with an Intel Core i7 configuration (16 GB RAM). The visual office scene, developed in Unity software with VBA framework was presented to the participants through HTC Vive headset [62] as shown in Figure **7.11**. In the audio-visual scene, the virtual classroom was visualized consisting of same furniture as in the real one. The same task with the same audio visual stimuli and timing settings was used in the real classroom experiment.

Figure 7.10: *Two trajectories presented from left to right and from right to left relative to the participant's orientation in front of the facade*

Figure 7.11: *HMD view point of the virtual receiving classroom and the experimental setup in VR environment*

The background sound conditions were auralized and played-back in such a way that a highest similarity of listening level between the two environments (i.e. real and VR) was provided. Similarly, the visual scene in VR is presented to the participants exactly the same as the real scene. Sound conditions for VR experiment were derived from the recordings of an accelerated motorbike engine sounds (at the speed of $40\frac{km}{h}$) in free-field conditions. The recorded sounds were auralized in the receiving VR classroom at a level of $L_{eq} = 40\text{dB}(A)$ resulted from auralizing the sound from outside and transmitted to the listener in the receiving classroom via façade.

In both experiments (real and VR), all background sounds were presented binaurally using Sennheiser HD 650 headphones. The sound pressure levels refer to an energy-equivalent sound pressure level L_{eq} averaged over presentation duration and measured using an arti cial ear (HMS-III, dummy head from HEAD Acoustics) and a sound level meter (Norsonic Sound Analyser 110) and calibrated the SPL for each sound condition through RME Fireface UC II sound card to deliver the target SPL to the headphones.

Figure 7.12: *HMD view of virtual classroom with experimental setup (a) GUI-Instruction, (b) GUI-Selection of position, (c) GUI-Answers and (d) GUI-Feedback*

The real experiment was conducted in classroom where as the VR experiment was conducted in VR lab at the Institute of Technical Acoustics (ITA) in individual sessions, whereas, the real experiment was conducted at real classroom (ITA "Seminar room (room number 011)" with closed window shades. It lasted approximately 10 min for one session. In the real classroom, the motorbike passed the building randomly 2 times from left to right and vice versa. In the virtual classroom, the written instructions were display in VR scene through HMD (Figure **7.12(a)**) and the participants were asked to ignore the background sounds from inside the building. At the beginning of the experiment, two practice trials were given under random sound conditions (i.e. play-back of sound source moving from either left to right or right to left). In each trial, the participant selects a listening position first in the room through a GUI on the HMD screen (Figure **7.12(b)**) and click on the "PLAY" pushbuttons for starting sound stimuli. Subsequently, sound is played back in random order (i.e. moving sound source from left to right or from right to left). The sound is played-back for 30 secs and after a pause of 5 seconds (retention interval) a GUI appears to ask the user for selecting correct answer, i.e. whether pass-by sound source was moving from left to right or from right to left as depicted in Figure **7.12(c)**.

The participants were expected to answer correctly by clicking on the pushbuttons using a virtual hand controller device. After clicking on one of the options the main GUI appeared again for selecting another position in the room. During each background sound condition, only two successive trials had to be completed in one experimental block. A succession of sound conditions was balanced over participants. Subjective ratings were collected after each background sound condition (Figure **7.12(d)**). Subjective ratings of "level of immersion", "difficulty in localization", "difficulty in listening level" and "hardness" were measured on a five-point scale with (**1**) = 'not at all', (**2**) = 'a little', (**3**) = 'middle', (**4**) = 'rather' and (**5**) = 'extremely'. Table X shows detail list of the questions asked at the end of experiment.

Table 7.1: *Subjective Rating Questionnaire*

		5	4	3	2	1
1.	How much effort put in listening level?	Extreme	Rather	Middle	A little	Not at all
2.	How much effort put in localizing source?	Extreme	Rather	Middle	A little	Not at all
3.	What was overall hardness of the test?	Extreme	Rather	Middle	A little	Not at all
4.	How realistic was the visual scene?	Extreme	Rather	Middle	A little	Not at all
5.	How realistic was the sound of siren?	Extreme	Rather	Middle	A little	Not at all

7.2.3.2.   Results

The localization of a moving source is a quite challenging task for the listener who is inside dwellings due to diffuse field of receiving rooms. Nevertheless, there is the precedence effect which leads to localization of sound source according the arrival of the first wave front. As the radiation from the secondary source are properly modelled concerning their relative group delays, the hypothesis can now be checked, whether or not the complex transmission and radiation situation can lead to localization effects in the room. It was expected that the listening experiment would be easy in localization of outdoor moving source inside the buildings for direct sound field. The reason for this assumption is that when only direct sound field is taken into account, the received signals on each façade element differ only by means of arrival time, directivity of the sound source and the energy level, hence, each façade element receives different amount of energy from the source and transmits it correspondingly inside the classroom. Furthermore, due to the angle dependencies of transmission coefficient of façade elements (on which the sound energy impinges at different angles of incidence from source) also transmit different sound energies inside the room.

The data collected in the present real and VR experiments were analysed and compared with each other. The goal was to compare the correct answers from the participants obtained in audio-visually rendered environment (VR) with those found in the real classroom environment. In the first step, it is evaluated that participants were able to perceptually localize the moving source under building-acoustical conditions in both real as well as virtual environments. Secondly, we compared both results to see if there is apparently a difference between the responses of participants in both environments. The results of this experiment are shown in Figure **7.13** for four different positions in the classroom (position 1 is close to the façade and position 4 is in the middle of the classroom, as shown in Figure **7.9**). Similarly, in the subjective ratings results in Figure **7.14**, we can see similar trends in both environments. The sounds were presented indicating different directions of the motorbike pass-by. In this preliminary test, in all cases the test subjects could identify the directions significantly above the guessing threshold of **50%**. It seems that in positions 2 and 3 this decision is easier than in position 1 (close to the façade) and position 4 (in the middle of the room). Due to the low number of tests, however, the results show only tendencies. More tests have to follow in order to meet the requirements of statistics. The tendency, however, is visible in the measured scenario as well as in the VR auralization. Nevertheless, the results seem to support the hypothesis that close to the façade (position 1), the main directional information is included on the strong direct sound of the closest patch (the window in view direction), so the sound source is localized just there. In contrast, further away, more windows contribute to the direct sound cluster with varying arrival times, so that the precedence effect and the windows with the earliest arrival time determines the source localization. At position 4, finally the diffuse field dominates, so that source localization gets more difficult. From the results of both studies the mean correct answers for all four positions are above 80% in both environments, hence it might be concluded that the perceptual localization of the human under sound insulations is high even though the source itself is visible. This study provides initial results of localization of moving outdoor sources which may lead to further experiments where more complex outdoor sound field can be taken into account an and continuing the studies on cognitive tests to find out the influence of the moving source and other intermittent outdoor noise stimuli on the performance of human during their daily work in built environments.

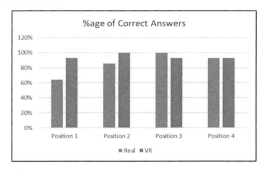

Figure 7.13: *Percentage of correct answers for the Real and VR experiments*

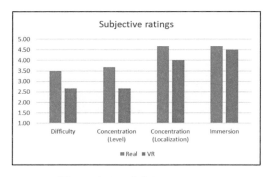

Figure 7.14: *Subjective ratings*

8

Summary

During the course of this thesis project, a large software framework for virtual building acoustics (VBA) was developed. The concept and the implementation of interactive real-time building acoustics simulations were described, which plays a vital role in interactive 3D audio rendering in virtual reality environments. The aim of this work was to establish an interface between psychoacoustic research and building acoustics in virtual built-up environments integrated with audio-visual virtual reality. Having introduced a brief background knowledge of fundamentals of room acoustics and basics of the building acoustics (sound insulation predictions), and the description of a concept for the multi-modal representation of indoor and outdoor sound sources, airborne sound insulation filter construction methods were introduced that took important sound propagation and transmission phenomenon into account. The sound insulation prediction methods were developed based on available standards and up-to-date research. The real-time performance was achieved by introducing VBA framework that enabled real-time modifications of sound sources and receivers. The main key features of VBA framework are described in Chapter **6**.

The extended sound insulation prediction models developed in this thesis are not solely based on assumptions of ideal diffuse sound fields rather based on room acoustics simulations which enables to calculate the energies at the surface of building elements for scenes differing from standard settings. This way, the sound insulation quantities were computed for any kind of source and receiver positions also in complex structure of the building elements. Certain challenges in the traditional work of sound insulation prediction methods were addressed from technical perspective and the improvements were made in sound insulation rendering with corresponding filter constructions for auralization. The building elements were considered as collection of coherent point sources rather than taking whole wall as single point source radiators and the bending wave patterns are addressed in order to construct the transfer functions from source to the receiver room. The room acoustical simulations are carried out for both source and receiving rooms to generate transfer functions from source to the source room walls and from radiating receiving room walls to the listener, so that the geometries and reverberation might be fit to the spatial impression of the listening rooms. In addition, the transfer functions from radiating walls of the receiving room to listener are designed in such a way that not only indoor sources are handled but outdoor moving sources are addressed as well.

Having achieved an advance sound insulation auralization framework, the accuracy of this framework, the quality of corresponding filters and verification of filter rendering was discussed. The claim of this work is not to achieve a perfect prediction result for an actual building which can be compared with real results obtained in on-site measurements. The verification was conducted by reproducing the results of D_{nT} from developed model are compared with that of ISO standards in terms of performance of the building elements. This was done for different indoor and outdoor scenes. The model was thus validated with reference to standard conditions produced by simulations based on ISO standards, where it was shown that the auralization results reproduce the sound insulation data with overall deviations of not more than 0.6 dB. The model was not validated in the sense that it was not compared to actual measurements for a situation as such a validation may be difficult for practical reasons. The real-time implementation was discussed and the latencies of the building acoustics auralization filter rendering are calculated. It was also shown that the developed sound insulation model might not only be used as prediction tool for sound insulation metrics for building elements but also provide the opportunity to construct filters for an interactive real-time auralization under different building acoustical conditions starting from a simple adjacent rooms to the complex urban scenes.

The building acoustic model was implemented in virtual reality environment concerning building acoustics aspects and auralization of sound insulation of indoor and outdoor spaces. We discussed basics of 3D graphics rendering tools that are vital for building acoustic implementation to create virtual architectural scenes and related audio rendering techniques such as; architectural design, geometry manipulation, real time convolution, digital signal processing and sound insulation filters rendering that make building acoustics virtual reality realistic and immersive. Above all, the main focus was to pay attention to the particular requirements of VR environments including performance issues and making sure that the sound insulation filters run fast enough in such environments. In Unity software the virtual building acoustics framework (VBA) was developed as package form including room acoustics, building acoustics, geometry handling and audio rendering tools. In this way, a universal platform for such kind of advanced virtual building acoustic frameworks was discussed. We also discussed auralization processing chains of the VBA framework, its evaluation and real-time performance with example audio-visual scenes. In Chapter **7**, it was discussed the application of the VBA framework for evaluation of the performance of the buildings and designing listening experiments such as evaluation of background noise impacts on cognitive performance of humans under different building acoustical conditions and effects of intermittent outdoor moving sound sources on perceptual localization. A complete open-source documentation is available on the VBA website, and few example case studies are presented at www.virtualbuildingacoustics.org [**61**].

9

Outlook

The contribution of this work is the development of a universal research platform to interface psychoacoustic research with building acoustics of virtual architectural environments and integration of the developed technology into audio-visual virtual reality systems. Though the virtual building acoustics (VAB) framework provides a good real-time performance of the whole building acoustic simulations from technical perspective, there are certainly many aspects of room and building acoustics that still are required to be investigated and integrated into this framework. This includes improvements of the applied room acoustics simulation models, as well as new simulation strategies and algorithmic concepts that further reduce the overall computation time. Therefore, some important ideas for future changes are discussed in the following paragraphs that would further motivate the scope of VBA application range.

One limitation of the auralization model is the precision of the angle-dependent sound transmission approach. For the time being, only monolithic building elements were modelled which sets prerequisites for the variety of building façade construction. In any case, however, the auralization model can be fed with experimental results. This way, the real-time auralization model is open for improvements in theory and experiment of building elements. Also, the auralization model could be "tuned" in optimization processes of matched input data for reference output data, such as in [74], in order to obtain a nominally "exact" model of an existing building, if that's required in specific applications.

In the room acoustics package, the source room impulse responses were synthesized at the surfaces of different walls to estimate the correct amount of sound energy, whereas in the receiver room the room impulse responses were synthesized to get the spatial impression of the room for auralization. Though the impulse response synthesis process in VBA could be based on IS and RT techniques, which are further required to be improved and validated, however, the main simulations so far are based on statistical approaches. It has to be checked up to which setting of image source order and ray tracing setting the auralization is still feasible in real time. Alternatively, the room acoustics filter part could be solved on a separate CPU.

In real dwellings, the rooms have irregular shapes and occupied with furniture with other absorptions. Therefore, geometrical approaches to calculate the valid room impulse responses for any arbitrary source position to the surfaces of the walls should be considered. In this way, the sound energy distribution on the walls might be calculated accurately and hence the sound transmission from specific paths could be precise. As the receiving room walls are segmented into multitude of secondary sources and hence for each individual secondary source the room impulse response and corresponding binaural transfer function were computed. However, the directivity patterns of radiation efficiencies of the secondary source radiators were not taken into account. By introducing the radiation patterns of the secondary sources and calculating the correct amount of energy flow from the secondary sources to the listener might increase the perception of loudness and localization capabilities of the actual sound source within the dwellings. Another approach might be to take vibration velocity distributions of the structures into account using radiation patterns of the whole vibrating walls to calculate the sound pressure level at the listener position. Another important aspect in normal dwellings is presence of slit sources due to very slight openings of the doors which significantly contributes to the level differences and coloration of the sound field. For this, modelling the slightly open doors/windows as slit sources certainly helps in calculating correct amount of energy propagating in the receiver rooms. It might increase the level of realism in auralization of building acoustic situations.

Outdoor sound propagation package in VBA includes acoustic simulation of urban sound propagation models which are helpful in many research and urban planning areas and might also be useful in early designing stages of such environments. These simulation models contribute to characterization of acoustic properties of these environments, especially for auralization of noise and evaluation of its effect in ecologically more valid manners. VBA includes VA software which takes into account important quantities and parameters that which characterize the outdoor sound field along with important wave phenomena such as reflections and diffractions. The models used to calculate sound propagation paths from a moving outdoor source to receiving points on a façade, utilizes the geometrical acoustics, which is based on the principle of replacing sound waves with sound particles travelling along a ray. VA has many real-time implementation issues and still needs to be validated for the accuracy of simulation results. It is therefore, required to design and implement a complete outdoor sound propagation model for real-time applications of façade sound insulations concerning the complex outdoor urban environments. Also here, it must be checked the limits of real-time performance. As discussed in Chapter **7**, the potential of using a human-centred approach integrating audio-video virtual reality to evaluate indoor noise and outdoor protection by building characteristics opens new ways of conducting listening experiments.

A

Annexes

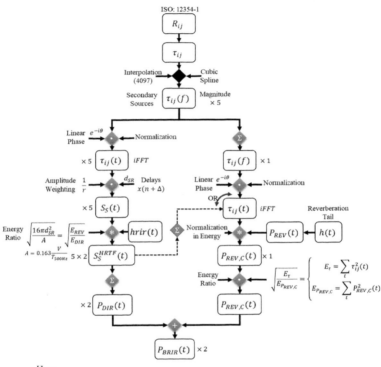

Here,

R_{ij} (τ_{ij}): Sound reduction (transmission) from element i to element j

S_s: Secondary source time domain signal

E_{rev} (E_{dir}): Energies of reverb (direct) sound field

d_{SR}: Distance between source and receiver

$hrir(t)$: Room impulse response

$S_s^{HRTF}(t)$: Binaural secondary source time domain signal

$P_{DIR}(t)$: Sound pressure (Direct Part)

$P_{REV}(t)$: Sound pressure (Reverb Part)

$P_{BRIR}(t)$: Binaural sound pressure

A.1: *Computational flow chart for sound insulation filters for adjacent rooms*

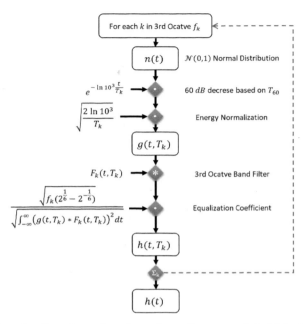

A.2: *Algorithm flow chart of synthesised reverberation tail for $h(t)$, from [64]*

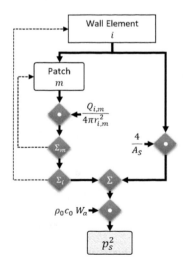

A.3: *Algorithm flow chart to calculate sound pressure at any wall patch from source*

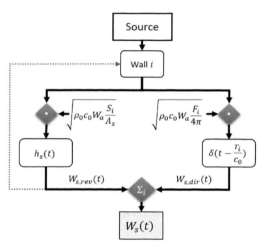

A.4: *Algorithm for computing energy (or sound pressure) at the surface of the source room wall element*

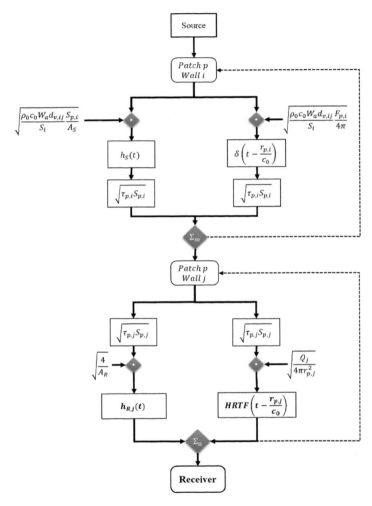

A.5: *Algorithm flow for Final binaural signal at receiving point*

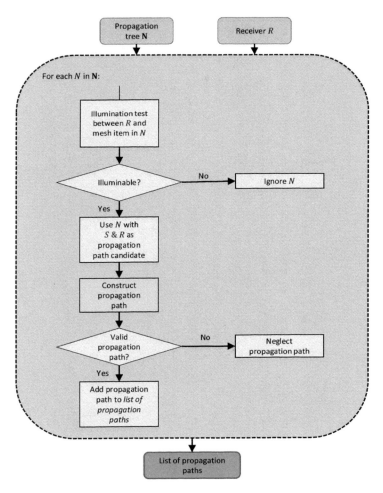

A.6: *Propagation path finding Algorithm*

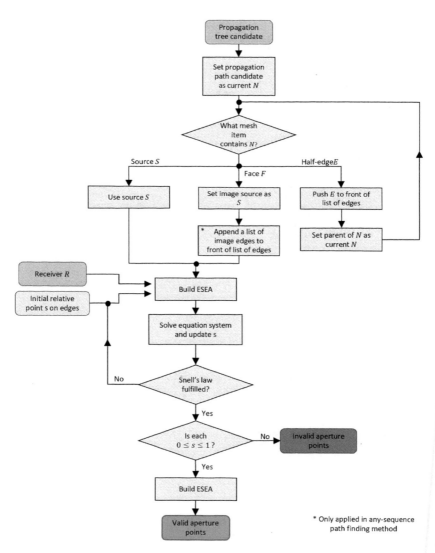

A.7: *Propagation path construction, reproduced from* **[67]**

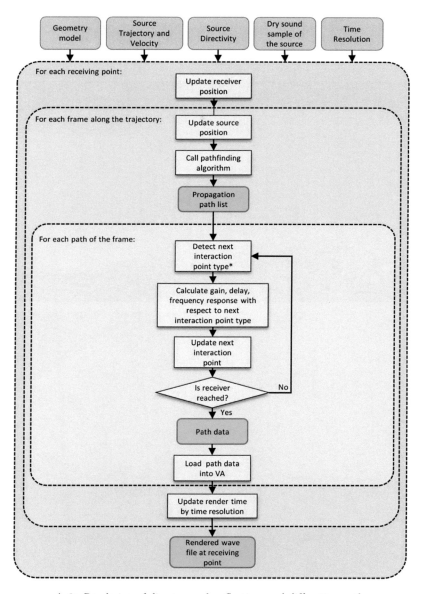

A.8: *Rendering of direct sound, reflection, and diffraction paths*

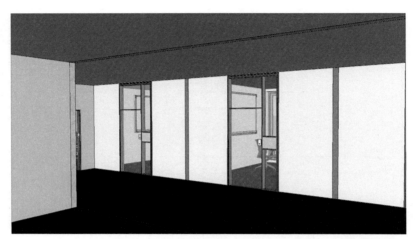

A.9: *ITA-Building: Internal view of architectural construction model*

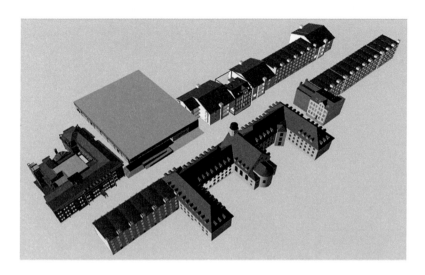

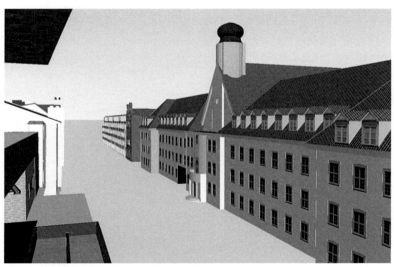

A.10: *Urban Environment Scene 1: External views of construction model in (The green floor building is selected as case study)*

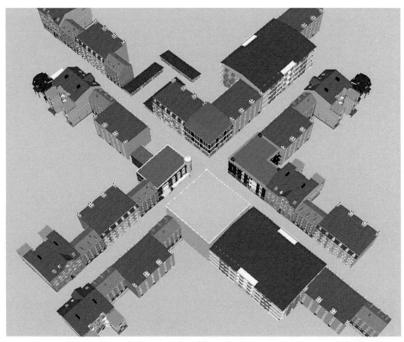

A.11: *Environment Scene 2: External views of construction model (The green celling building is selected as case study)*

A-12: *ITA building: Internal views in virtual reality*

A-13: *Urban Scene 1: External view of street canyon*

A-14: *Urban Scene 1: Internal view of selected receiving room*

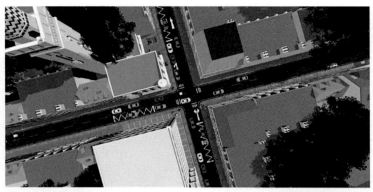

A-15: *Urban Scene 2: External view of a crossroad junction*

A-16: *Urban Scene 2: Internal view of selected receiving room*

A-17: *Audio-Visual Scenes*

Bibliography

[1]. Hongisto, V. J., Leppämäki, H., Oliva, D. and Jukka H. "Work performance in private office rooms: The effects of sound insulation and sound masking." Building and Environment 104: 263-274. (2016)

[2]. Liebl, A. and Jahncke, H. "Review of research on the effects of noise on cognitive performance 2014-2017." In 12th ICBEN conference on noise as a public health problem, Zurich, Switzerland. (2017)

[3]. Thaden, R., "Auralization in Building Acoustics." PhD dissertation, RWTH University Aachen: Germany. (2007)

[4]. Rasmussen, B. "Harmonization of sound insulation descriptors and classification schemes in Europe: COST Action TU0901," in Proceedings of European Symposium on Harmonisation of European Sound Insulation Descriptors and Classification Standards, Florence. (2010): https://www.cost.eu/actions/TU0901

[5]. Schlittmeier, S. J., Hellbrück, J., Thaden, R. and Vorländer, M. "The impact of background speech varying in intelligibility: Effects on cognitive performance and perceived disturbance," Ergonomics 51: 719–736. (2008)

[6]. ISO: 12354-1, "Building acoustics: estimation of acoustic performance of buildings from the performance of elements, Part 1: Airborne sound insulation between rooms.", European Committee for Standardization Brussels. (2017)

[7]. Vorländer, M. and Imran M., "Real-time auralization of sound insulation, in Inter Noise-2018, Illionis: Chicago." (2018)

[8]. Muhammad, I., Vorländer M., and Schlittmeier S. J. "Audio-video virtual reality environments in building acoustics: An exemplary study reproducing performance results and subjective ratings of a laboratory listening experiment." The Journal of the Acoustical Society of America. 146(3): EL310-6. (2019)

[9]. Vorländer, M., "Auralization: Fundamentals of Acoustics, Modelling, Simulation, Algorithms and Acoustic Virtual Reality." Springer. (2007)

[10]. Vorländer, M. "Building acoustics: from prediction models to auralization." In Proceedings of Acoustics. (2006)

[11]. Zienkiewicz, O. C., et. al., "The finite element method." Vol. 3. McGraw-Hill London. (1977)

[12]. Reynders E., Langley R. S., Dijckmans A. and Vermeir G. "A hybrid finite element–statistical energy analysis approach to robust sound transmission modelling." Journal of Sound and Vibration. 333(19):4621-36. (2014)

[13]. Lyon, R. H. and DeJong, R. G. "Theory and Application of Statistical Energy Analysis." Ed. 5. USA: Butterworth-Heinemann. (1995)

[14]. Vorländer, M. and Thaden R. "Auralization of Airborne Sound Insulation in Buildings." Acustica united with Acta Acustica 86: p. 7. (2000)

[15]. Hopkins, C. "Sound insulation." Routledge. (2015)

[16]. ISO: DIN-717-1: "Acoustics-Rating of sound insulation in buildings of building elements - Part I. 2006.1." (1996)

[17]. ISO: 12354-3, "Building acoustics: estimation of acoustic performance of buildings from the performance of elements, Part 3: Airborne sound insulation against outdoor sound.", European Committee for Standardization Brussels. (2017)

[18]. Gerretsen, E., "Calculation of the sound transmission between dwellings by partitions and flanking structures." Journal of Applied Acoustics, 12(6): p. 413-433. (1979)

[19]. Fiebig, A. "The perception of acoustic environments and how humans form overall noise assessments." In Internoise Congress and Conference Proceedings. Vol. 259. No. 1., Institute of Noise Control Engineering. (2019)

[20]. Vigran, T. Erik. "Building acoustics." CRC Press, (2014)

[21]. Kuttruff, H. "Room Acoustics." 4 Ed. Taylor & Francis. (2001)

[22]. Rindel, J. H. "Sound Insulation in Buildings.", U.S: CRC Press. (2018)

[23]. Fahy, F. J. and Gardonio, P. "Sound and structural vibration: radiation, transmission and response." Elsevier. (2007)

[24]. Villot, M., Guigou, C. and Gagliardini, L. "Predicting the acoustical radiation of finite size multi-layered structures by applying spatial windowing on infinite structures." Journal of sound and vibration 245, No. 3: 433-455. (2001)

[25]. Davy, J. L. "Predicting the sound insulation of single leaf walls: Extension of Cremer's model." The Journal of the Acoustical Society of America 126, No. 4: 1871-1877. (2009)

[26]. Crocker, M. J. and Price, A. J. "Sound transmission using statistical energy analysis." Journal of Sound and Vibration 9, No. 3: 469-486. (1969)

[27]. Sewell, E. C. "Transmission of reverberant sound through a single-leaf partition surrounded by an infinite rigid baffle." Journal of Sound and Vibration 12, No. 1: 21-32. (1970)

[28]. DIN 4109-2, "Sound insulation in buildings, Part 2: Verification of compliance with the requirements by calculation." (2018)

[29]. Vigran, T. E. "Predicting the sound reduction index of finite size specimen by a simplified spatial windowing technique." Journal of Sound and Vibration 325, No. 3:507-512. (2009)

[30]. Ryzhik, I. M. and Gradshtein, I. S. "Table of integrals, series, and products.", New York: Academic Press. (1965)

[31]. Cremer, L. "Theorie der Schalldämmung von Wänden bei schrägem Einfall, Akustische Zeitschrift, 7, 81–104. (1942).

[32]. Mulholland, K.A., Parbrook, H.D. and Cummings, A. "The transmission loss of double panels." Journal of Sound and Vibration. Nov, 1:6(3):324-34. (1967)

[33]. Bolton, J. S., Shiau, N. M. and Kang, Y. J. "Sound transmission through multi-panel structures lined with elastic porous materials." Journal of Sound and Vibration., 4, 191(3):317-47. (1996)

[34]. Sato, H. "Transmission of traffic noise through windows Influence of incident angle on sound insulation in theory and experiments." Journal of the Acoustical Society of Japan, 29, 509-516. (1973)

[35]. Rindel, J. H. "On the mechanism of outdoor noise transmission through walls and windows." The Acoustics Laboratory, Technical University of Denmark, Report No. 9. (1975)

[36]. Ljunggren, S. "Airborne sound insulation of thin walls." Journal of the Acoustical Society of America, 89, 2324-2337. (1991)

[37]. Sewell, E. C. "Transmission of reverberant sound through a single-leaf partition surrounded by an infinite rigid baffle." Journal of Sound and Vibration, 21-32. (1970)

[38]. Leppington, F. G. et al. "Resonant and non-resonant acoustic properties of elastic panels. I. The radiation problem." In proceedings of the Royal Society of London, A 406, 139-171. (1986)

[39]. Leppington, F. G. et al. "Resonant and non-resonant acoustic properties of elastic panels. II. The transmission problem."" In proceedings of the Royal Society of London, A 412, 309-337. (1986)

[40]. Cremer, L., Heckl, M. and Petersson, B.A.T. "Structure-Borne Sound: Structural Vibrations and Sound Radiation at Audio Frequencies." Springer (2005)

[41]. Nightingale, T. R. T., and I. Bosmans. "Expressions for first-order flanking paths in homogeneous isotropic and lightly damped buildings." Acta Acustica united with Acustica 89, No. 1: 110-122. (2003)

[42]. Gerretsen, E. "Calculation of airborne and impact sound insulation between dwellings." Applied Acoustics. 19(4):245-64. (1986)

[43]. Wenmaekers, R.H.C., Hak, C.C.J.M., Hornikx, M.C.J. and Kohlrausch, A.G., "Sensitivity of stage acoustic parameters to source and receiver directivity: Measurements on three stages and in two orchestra pits." Applied Acoustics, 123, pp.20-28. (2017)

[44]. Vorländer, M. "Revised relation between the sound power and the average sound pressure level in rooms and consequences for acoustic measurements." Acta Acustica united with Acustica 81, No. 4: 332-343. (1995)

[45]. Hassan, O. A. "Building acoustics and vibration: theory and practice." World Scientific Publishing Company. (2009)

[46]. Imran, M., Vorländer M. and Heimes, A. "Auralization of Airborne Sound Transmission and Framework for Sound Insulation Filter Rendering" In Proceeding of Euronoise 2018. (2018)

[47]. Imran, M., Heimes, A. and Vorländer, M. "A new approach for real-time sound insulation filters development." In Internoise-2019, Spain. (2019)

[48]. Rasmussen, B. "Optimizing the sound insulation of the normal double glazing" Lydteknisk Institut, Rapport No: 13. (1984)

[49]. Hongisto, V., Lindgren, M. and Helenius, R. "Sound insulation of double walls - an experimental study." Acta Acustica United with Acustica 88:904–923. (2002)

[50]. Ljunggren, S. "Airborne sound insulation of thick walls," J. Acoust. Soc. Am. 89: 2338–2345. (1991)

[51]. Northwood, T. D. "Transmission loss of plasterboard walls," Building Research Note, BRN-66, Division of Building Research, National Research Council of Canada, Ottawa. (1968)

[52]. Lentz T, Schröder D, Vorländer M, Assenmacher I. "Virtual reality system with integrated sound field simulation and reproduction." EURASIP journal on advances in signal processing. (1):070540. (2007)

[53]. Vorländer, M., Schröder, D. Pelzer, S. Wefers, F. "Virtual reality for architectural acoustics." Journal of Building Performance Simulation. 8(1):15-25. (2015)

[54]. Schröder, D. "Physically based real-time auralization of interactive virtual environments." PhD dissertation, RWTH University Aachen: Germany. (2011)

[55]. Huopaniemi, J., Savioja, L. and Takala, T. "DIVA virtual audio reality system." Georgia Institute of Technology.

[56]. Klatte, M., Lachmann, T., Schlittmeier, S. and Hellbrück, J. "The irrelevant sound effect in short-term memory: Is there developmental change?" European Journal of Cognitive Psychology. (22), 1168-1191. (2010)

[57]. Schlittmeier, S. J. and Hellbrück, J. "Background music as noise abatement in open plan offices: A laboratory study on performance effects and subjective preferences," Applied Cognitive Psychology. 23, 684-697. (2009)

[58]. Keus, P. M., Carlsson, J., Marsh, J. E., Ljung, R., Odelius, J., Schlittmeier, S. J. Sundin, G. and Sörqvist, P. "Unmasking the effects of masking on performance: The potential of multiple-voice masking in the office environment," J. Acoust. Soc. Am. 138, 807-816. (2015)

[59]. Renz, T., Leistner, P. and Liebl, A. "Auditory distraction by speech: Can a babble masker restore working memory performance and subjective perception to baseline?" Applied Acoustics. 137, 151-160. (2018)

[60]. http://virtualacoustics.org

[61]. http://virtualbuildingacoustics.org

[62]. https://vive.com

[63]. https://unity.com

[64]. Rodríguez-Molares, A. "A new method for auralization of airborne sound insulation." Applied Acoustics, 74(1): 116-121. (2013)

[65]. Imran, M., Heimes A., and Vorländer, M. "Sound insulation auralization filters design for outdoor moving sources." In proceedings of 23rd International Congress on Acoustics, Aachen, 283-288. (2019)

[66]. Imran, M., Heimes, A. and Vorländer, M. "Perceptual Localization in Virtual Reality Environments of Pass-by Outdoor Sources under Sound Insulation Conditions." In proceeding of DAGA-2020, Aachen, Germany. (2020)

[67]. Erraji, A., Stienen, J. and Vorländer, M. "The image edge model. Acta Acustica, 5: 17. (2021)

[68]. Vanekckek, G. Jr., "Back-face culling applied to collision detection of polyhedra", The Journal of Visualization and Computer Animation, Vol. 5, No. 1. (1994)

[69]. ISO 10140-2: "Acoustics: Acoustics-Laboratory measurement of sound insulation of building elements - Part 2: Measurement of airborne sound insulation." International Organization for Standardization, Geneva (2010).

[70]. Wefers F. "Partitioned convolution algorithms for real-time auralization." PhD dissertation, RWTH University Aachen: Germany. (2015)

[71]. http://sketchup.com

[72]. http://datakustik.com

[73]. Meng, F. "Modelling of Moving Sound Sources based on Array Measurements." PhD dissertation, RWTH University Aachen: Germany. (2018)

[74]. Aspock, L. "Validation of Room Acoustics simulation models." PhD dissertation, RWTH University Aachen: Germany. (2020)

[75]. https://www.vive.com/product

[76]. Välimäki, V., Pakarinen, J., Erkut, C., Karjalainen, M. "Discrete-time modelling of musical instruments." In Reports on progress in physics. 69(1):1. (2005)

[77]. https://odeon.dk

[78]. Reynders, E. P. B., Wang, P., Hoorickx, C. V., and Lombaert G. "Prediction and uncertainty quantification of structure-borne sound radiation into a diffuse field.", Journal of Sound and Vibration, Volume 463. (2019)

[79]. Craik, R. J. M. and Smith, R. S. "Sound transmission through Double Leaf Lightweight Partitions Part-I, Airborne Sound.", Applied Acoustics, 61(2). (2000)

[80]. Muhammad, I., Heimes, A. and Vorländer, M. "Interactive real-time auralization of airborne sound insulation in buildings." Acta Acustica, 5, 19. (2021)

[81]. https://www.fftw.org

[82]. Heimes, A., "Filter Design for Sound Insulation Auralization." Master Thesis, RWTH University Aachen, Germany. (2019)

[83]. Waterhouse, R. V., "Interference Patterns in Reverberant Sound Fields." J. Acoust. Soc. Am. 27: 247-258. (1955)

[84]. Ridwan-ur-Rehman, M., "Effects of sound in buildings on the human cognitive performance." Master Thesis, RWTH University Aachen, Germany. (2019)

[85]. ISO: 15186-1, "Acoustics - Measurement of sound insulation in buildings and of building elements using sound intensity - Part 1: Laboratory measurements.", European Committee for Standardization. (2000) (Revised: 2021)

[86]. ISO: 15186-2, "Acoustics - Measurement of sound insulation in buildings and of building elements using sound intensity - Part 2: Field measurements.", European Committee for Standardization. (2000) Revised: 2020)

Curriculum Vitae

IMRAN Muhammad
M.Sc. (Physics)
Research Assistant (Wissenschaftl. Mitarbeiter)

Institute of Technical Acoustics,
RWTH Aachen University
Kopernikusstr. 5
52074 Aachen, Germany
📞 +49 241 80 979 98 | +49 176 433 83516
✉ mim@akustik.rwth-aachen.de

EDUCATION	❖ **M.Sc. {Master's Physics}** from University of the Punjab, Lahore Pakistan ❖ **B.Sc. {Graduation} (Pure Mathematics, Applied Mathematics, Physics)** from University of the Punjab, Lahore Pakistan
RESEARCH PROJECTS	**THE OUTLINES OF THE RESEARCH PROJECTS:** {2004 TO 2016} ➢ Project on "Building-acoustic Auralization Test Environment for Psychoacoustic Experiments with Contextual and Interactive Features" **{2017-2021}** ➢ Project on "Real time Virtual Spatial Sound Rendering for VR" **{2015-2016}** ➢ Project on "Development of Real Time Methods for Rendering Dynamic/Interactive 3D Virtual Sound for Telepresence and Coexistence Virtual reality environment" under **CHIC, KIST** (http://chic.re.kr/eng/) **{2014-2016}** ➢ Design and development of Spherical Microphone Arrays for 3D sound analysis, synthesis and visualization **{2014-2015}** ➢ Real time Localization, Tracking and Beamforming methods using microphone arrays for speech processing and synthesis for robots **{2013-2015}** ➢ Evaluation and Computation of Sound Field Diffuseness in Opera House, Concert Hall and Orchestra **{2013-2014}** ➢ High Speed Train Noise (KTX-Korea) Measurement and Design Solutions **{2013}**

	➢ Design and development of Sonic Detection and Ranging (SODAR) System. The ability of the System is to localize and Track Sound Source {**2007-2012**}
EMPLOYMENT (**Work Experience**)	❖ **Institute of Technical Acoustics** (RWTH Aachen University) (**April-2017 to April-2021**) **Job Description:** Research Assistant/Teaching • Acoustic Virtual Reality • Room and Building Acoustics ❖ **Architectural Acoustics Lab** {Hanyang University, Korea} (**March-2013 to March-2016**) **Job Description:** Research Assistant • Room Acoustics, Architectural Acoustics (Spatial-temporal analysis of sound field) • Acoustic for Virtual Reality • 3D sound reproduction for telepresence and Virtual Reality • Microphone Array design and development based Smart Sensor, Localization, tracking of acoustic sources and Beamforming Techniques ❖ **CESAT,** {NESCOM Islamabad, Pakistan} (**Jan-2004 to Feb-2017**) **Job Description:** Research Assistant • Microphone arrays: Design, development and Beamforming Techniques • Developed "A sound detection and ranging system (SODAR®)" • Developed "Absolute Gravity Model for Pakistan (PGM®)"

Bisher erschienene Bände der Reihe

Aachener Beiträge zur Akustik

ISSN 1866-3052
ISSN 2512-6008 (seit Band 28)

Alle erschienenen Bücher können unter der angegebenen ISBN-Nummer direkt online (http://www.logos-verlag.de) oder per Fax (030 - 42 85 10 92) beim Logos Verlag Berlin bestellt werden.